Laboratory biosafety manual

Second edition

Laboratory biosafety manual

Second edition

World Health Organization
Geneva 1993

WHO Library Cataloguing in Publication Data

Laboratory biosafety manual. — 2nd ed.

    1.Containment of biohazards   2.Laboratories – standards
    3.Laboratory infection – prevention & control

    ISBN 92 4 154450 3   (NLM Classification: QY 25)

TYPESET IN INDIA
PRINTED IN ENGLAND

92/9365—Macmillan/Clays—9000

Contents

Foreword   ix

Acknowledgements   xi

1. General principles   1

Part 1. Guidelines   5

---

2. Basic laboratories – Biosafety Levels 1 and 2   7
Code of practice   7
Laboratory design and facilities   10
Laboratory equipment   12
Health and medical surveillance   13
Training   14
Decontamination and disposal   15
Chemical, fire, electrical and radiation safety   18

3. The Containment laboratory – Biosafety Level 3   19
Code of practice   19
Laboratory design and facilities   20
Laboratory equipment   21
Health and medical surveillance   21

4. The Maximum Containment laboratory – Biosafety Level 4 24
Laboratory design and facilities   24

5. Laboratory animal facilities   27
Animal facility – Biosafety Level 1   28
Animal facility – Biosafety Level 2   28
Animal facility – Biosafety Level 3   29
Animal facility – Biosafety Level 4   30
Invertebrates   30

Part 2. Good microbiological technique   33

6. Safe laboratory techniques   35
Techniques for the safe handling of specimens in the
    laboratory   35
Techniques for the use of pipettes and pipetting aids   36
Techniques for avoiding the dispersal of infectious
    materials   37
Techniques for the use of biological safety cabinets   37
Techniques for avoiding ingestion of infectious materials
    and their contact with skin and eyes   38
Techniques for avoiding injection of infectious materials   38
Techniques for the separation of serum   38
Techniques for the use of centrifuges   39
Techniques for the use of homogenizers, shakers and
    sonicators   40
Techniques for the use of tissue grinders (Griffith's
    tubes or TenBroek grinders)   41
Techniques for the care and use of refrigerators and
    freezers   42
Techniques for the opening of ampoules containing
    lyophilized infectious materials   43
Storage of ampoules containing infectious materials   43
Special precautions with blood and other body fluids   44
Precautions with materials that may contain
    "unconventional" agents   46

7. Safe shipment of specimens and infectious materials   48
Definitions   49
Documentation and packaging requirements   49
Transport-associated accidents: response and
    emergency safety measures   52

CONTENTS

8. Contingency plans and emergency procedures   55
Contingency plans   55
General emergency plans for microbiological laboratories   56

9. Disinfection and sterilization   60
Chemical disinfectants   60
Space and surface decontamination   63
Decontamination of biological safety cabinets   65
Sterilization   65
Incineration   69
Final disposal   70

Part 3. Laboratory equipment   71

10. Equipment-related hazards   73
Equipment that may create a hazard   73

11. Equipment designed to eliminate or reduce hazards   76
Biological safety cabinets   78
Pipetting aids   82
Homogenizers and sonicators   82
Disposable transfer loops   83
Microincinerators   83

Part 4. Chemical, fire and electrical safety   85

12. Hazardous chemicals   87
Storage of chemicals   87
Toxic effects of chemicals   89
Explosive chemicals   92
Chemical spillage   92
Compressed and liquefied gases   93

13. Fire in the laboratory   94

14. Electrical hazards   96

Part 5.   Safety organization and training   97

15.   The safety officer and safety committee   99
       Biosafety officer   99
       Biosafety committee   100
       General organization   101

16.   Safety rules for support staff   102
       Engineering and building maintenance services   102
       Cleaning (domestic) services   103

17.   Training programmes   104
       Basic course: Good laboratory practice   104
       Module 1 (the core module):
           Good microbiological technique   105
       Module 2: The safe laboratory environment   106
       Module 3: For support staff   107
       Module 4: For safety staff   108
       Module 5: For specialist staff who handle microorganisms
           in Risk Groups 3 and 4   109

Part 6.   Safety checklist   111

18.   Safety checklist   113
       Laboratory premises   113
       Storage facilities   113
       Sanitation and staff facilities   114
       Heating and ventilation   114
       Lighting   114
       Services   114
       Security   115
       Fire prevention   115
       Flammable liquid storage   115
       Electrical hazards   116
       Compressed and liquefied gases   116
       Personal protection   117
       Health and safety of staff   117
       Laboratory equipment   118
       Infectious materials   118
       Chemicals and radioactive substances   119

CONTENTS

References   120

Annex 1. National guidelines and codes of practice   123

Annex 2. Immunization of staff   126

Annex 3. Safety in microbiology: training information   127

Index   129

Foreword

The staff of microbiology laboratories, by definition, work with infectious organisms, or materials that do or may contain them. Some of these organisms are pathogenic or potentially so, depending on circumstances and dosage. Avoidance of infection is thus an essential element of the professional expertise of the workers. Safety is good technique—a hallmark of technical excellence in which the staff should take pride. It is necessary to protect not only the microbiologists themselves and their assistants and contacts, but also to protect their materials from possible cross-contamination which may invalidate their work by giving false results.

The first edition of this manual dealt primarily with these aspects of biosafety, but reminded readers of the additional risks and the need for consideration of chemical, physical and radiation hazards. In this second edition, this emphasis is rightly retained, but there is fuller treatment of the dangers and safety considerations relevant to fire, electrical apparatus and hazardous chemicals. The text has been updated in the light of experience, new discoveries and developments since the first edition. It gives useful technical detail covering laboratory equipment, including automated apparatus, disposal and sterilization.

The general guidance is based on the consensus of a group of experts. It is essentially practical, recognizing the variations in importance of particular pathogens and procedures in different countries, and in the available facilities and resources, levels of education and training of personnel. There is much valuable information for those who are developing arrangements for safe practices where these are not already or not fully established. Annex 1 includes examples of safety codes published in many countries, useful for those developing or reviewing their own practices and arrangements. This annex, together with the comprehensive reference list, enables the text to be concise so that, despite the amount of new information included, the manual remains conveniently compact for use as a practical handbook.

The six main parts of the manual deal respectively with:

— guidelines on basic laboratory design, equipment and operation for different levels of biosafety, including animal facilities;
— good microbiological technique and procedures;
— laboratory equipment and its use;
— chemical, fire and electrical safety;
— safety organization and training; and
— safety checklist—this is useful for those reviewing existing arrangements as well as for those planning to introduce new ones.

The detailed emphasis on education and training of staff is important. No biosafety cabinet or other facility or procedure alone guarantees safety unless the users operate safe techniques based on informed understanding. Safety equipment may even generate a false sense of security and carelessness leading to increased risk unless it is designed, installed, maintained and operated with proper understanding. Alert and intelligent self-monitoring during procedures is required, as well as appropriate supervision and surveillance.

At the same time a sense of proportion is necessary. Absolute safety is unattainable in any area of life or work. Well run laboratories are not specially dangerous workplaces, and one should avoid "overkill" from disproportionately stringent precautions or excessive reliance on physical safeguards. I endorse this caution given by Dr J. E. M. Whitehead in his introduction to the first edition of this manual, and I fully support his statement that "the principal element in biosafety is the inculcation of sound microbiological practices in both microbiologists and non-microbiologists". Use of this manual should help to achieve this desirable objective.

N. R. Grist
Emeritus Professor of Infectious Diseases,
University of Glasgow
former Head of Regional Virus Laboratory,
Ruchill Hospital, Glasgow, Scotland

Acknowledgements

The development of this manual has been made possible thanks to the contributions of the following scientists and safety officers. The people whose names are marked with an asterisk constituted an ad hoc editorial committee.

Dr C. H. Collins,* The Ashes, Hadlow, Kent, England (*Coordinating Editor*)

Dr S. G. Drozdov,* Institute of Poliomyelitis and Viral Encephalitides, Moscow, Russian Federation

Mrs M. E. Kennedy,* Division of Biosafety, Laboratory Centre for Disease Control, Ottawa, Ontario, Canada

Dr R. W. McKinney, Division of Safety, National Institutes of Health, Bethesda, MD, USA

Dr R. Möllby, National Bacteriological Laboratory, Stockholm, Sweden

Mr V. R. Oviatt,* Occupational Hygiene Unit, Wolfson Institute of Occupational Health, Dundee University Medical School, Ninewells, Dundee, Scotland

Dr Y. Pervikov,* Medical Officer, Microbiology and Immunology Support Services, Division of Communicable Diseases, World Health Organization, Geneva, Switzerland (*Secretary*)

Appreciation is also expressed to all those who helped in the production of the manual. They include, in addition to WHO staff members, the following:

Professor N. R. Grist, Emeritus Professor of Infectious Diseases, University of Glasgow, Glasgow, Scotland

Mrs M. Kennett, Virology Department, Fairfield Hospital, Victoria, Australia

Dr R. M. Rowan, Department of Haematology, Western Infirmary, Glasgow, Scotland

The World Health Organization thanks the National Institutes of Health, Bethesda, MD, USA, for their generous contribution to the preparation of the manual.

1. General principles

Throughout this manual, references are made to the relative hazards of infective microorganisms by risk group (Risk Groups 1, 2, 3, and 4). Laboratories are designated according to their design features, construction and containment facilities (safety precautions and equipment) as Basic – Biosafety Level 1, Basic – Biosafety Level 2, Containment – Biosafety Level 3, and Maximum Containment – Biosafety Level 4.

Table 1 describes the risk groups and Table 2 relates them to the laboratory designations. Table 3 gives a summary of the four biosafety level requirements.

Table 1. Classification of infective microorganisms by risk group

Risk Group 1 (no or very low individual and community risk)

A microorganism that is unlikely to cause human or animal disease.

Risk Group 2 (moderate individual risk, low community risk)

A pathogen that can cause human or animal disease but is unlikely to be a serious hazard to laboratory workers, the community, livestock or the environment. Laboratory exposures may cause serious infection, but effective treatment and preventive measures are available and the risk of spread of infection is limited.

Risk Group 3 (high individual risk, low community risk)

A pathogen that usually causes serious human or animal disease but does not ordinarily spread from one infected individual to another. Effective treatment and preventive measures are available.

Risk Group 4 (high individual and community risk)

A pathogen that usually causes serious human or animal disease and that can be readily transmitted from one individual to another, directly or indirectly. Effective treatment and preventive measures are not usually available.

1

Table 2. Relationship of risk groups to biosafety levels, practices and equipment

Risk group	Biosafety level	Examples of laboratories	Laboratory practices	Safety equipment
1	Basic – Biosafety Level 1	Basic teaching	GMT[a]	None; open bench work
2	Basic – Biosafety Level 2	Primary health services; primary level hospital; diagnostic, teaching and public health	GMT plus protective clothing; biohazard sign	Open bench plus BSC[b] for potential aerosols
3	Containment – Biosafety Level 3	Special diagnostic	As level 2 plus special clothing, controlled access, directional air flow	BSC and/or other primary containment for all activities
4	Maximum Containment– Biosafety Level 4	Dangerous pathogen units	As level 3 plus airlock entry, shower exit, special waste disposal	Class III BSC or positive pressure suits, double-ended autoclave, filtered air

[a] GMT, good microbiological technique.
[b] BSC, biological safety cabinet.

Each country should draw up a classification by risk group of the microorganisms encountered within its boundaries, based on the following factors:

— Pathogenicity of the organism.
— Mode of transmission and host range of the organism. These may be influenced by existing levels of immunity, density and movement of the host population, presence of appropriate vectors and standards of environmental hygiene.
— Availability of effective preventive measures. These may include: prophylaxis by immunization or administration of antisera; sanitary measures, e.g., food and water hygiene; control of animal reservoirs or arthropod vectors; and restrictions on the importation of potentially infected animals or animal products.
— Availability of effective treatment. This includes passive immunization, postexposure vaccination, and use of antimicrobials and chemotherapeutic agents, taking into consideration the possibility of the emergence of resistant strains.

2

Table 3. Summary of biosafety level requirements

	Biosafety level			
	1	2	3	4
Isolation of laboratory	No	No	Desirable	Yes
Room sealable for decontamination	No	No	Yes	Yes
Ventilation:				
inward air flow	No	Desirable	Yes	Yes
mechanical via building system	No	Desirable	Desirable	No
mechanical, independent	No	No	Desirable	Yes
filtered air exhaust	No	No	Yes	Yes
Double-door entry	No	No	Yes	Yes
Airlock	No	No	No	Yes
Airlock with shower	No	No	No	Yes
Effluent treatment	No	No	No	Yes
Autoclave:				
on site	Yes	Yes	Yes	Yes
in laboratory room	No	No	Yes	Yes
double-ended	No	No	Desirable	Yes
Biological safety cabinets				
Class I or II	No	Yes	Yes	Desirable
Class III	No	No	Desirable	Yes

In assessing the various criteria for classification it is also important to take into account conditions prevailing in the geographical area in which the microorganisms are handled. Individual governments may decide to prohibit the handling or importation of certain pathogens except for diagnostic purposes.

In the preparation of lists it is recommended that, where appropriate, some additional information is given about the advisability of wearing gloves and eye protection and, in the case of some Risk Group 3 pathogens, whether or not a biological safety cabinet is required.

Existing classifications, made by different countries and official organizations, may be useful for the preparation of new classifications and guidelines. Some of these are listed in Annex 1.

No special mention is made in this manual of work with genetically engineered microorganisms. These may be placed in the risk groups that are appropriate for their recipients and donors, and handled at the relevant biosafety level. Various national and international codes and guidelines for work with genetically engineered organisms are also included in Annex 1.

3

PART 1

Guidelines

2. Basic laboratories – Biosafety Levels 1 and 2

For the purposes of this manual, guidance and recommendations pertaining to Basic laboratories – Biosafety Levels 1 and 2 are directed at microorganisms in both Risk Group 1 and Risk Group 2. Although some of the precautions may appear to be unnecessary for some organisms in Risk Group 1, they are desirable for training purposes to promote good (i.e., safe) microbiological technique (GMT).

As no clinical or hospital laboratory has complete control over the specimens it receives, workers may occasionally and unexpectedly be exposed to organisms in higher risk groups. This possibility must be recognized in the development of safety plans and policies.

The Biosafety Level 2 guidelines presented here are comprehensive and detailed as they are fundamental to all classes of laboratories. The guidelines for the Containment laboratory – Biosafety Level 3 and the Maximum Containment laboratory – Biosafety Level 4 that follow later are modifications of these guidelines designed for work with the more dangerous pathogens.

Code of practice

This code is a listing of the most essential laboratory procedures that are basic to good microbiological technique. In many laboratories and national laboratory programmes, such a code may be given the status of "rules" for laboratory operation. In these guidelines, and elsewhere in this manual, various parts of the code of practice are elaborated and explained.

It is emphasized that good microbiological technique is fundamental to laboratory safety and cannot be replaced by specialized equipment, which can only supplement it.

The most important rules are listed below, not necessarily in order of importance.

7

1. The international biohazard sign (Fig. 1) should be displayed on the doors of rooms where Risk Group 2 microorganisms are handled.
2. Pipetting by mouth should be prohibited.
3. Eating, drinking, smoking, storing food and applying cosmetics must not be permitted in the laboratory work areas.
4. Labels must not be licked; materials must not be placed in the mouth.
5. The laboratory should be kept neat, clean and free of materials that are not pertinent to the work.
6. Work surfaces must be decontaminated after any spill of potentially dangerous material and at the end of the working day.
7. Members of the staff must wash their hands after handling infectious materials and animals, and before they leave the laboratory.

ADMITTANCE TO AUTHORIZED PERSONNEL ONLY

Hazard identity: _____

Responsible Investigator: _____

In case of emergency call: _____

Daytime phone _____ **Home phone** _____

Authorization for entrance must be obtained from
the Responsible Investigator named above.

Fig. 1. Hazard warning sign for laboratory doors

8. All technical procedures should be performed in a way that minimizes the formation of aerosols and droplets.

9. All contaminated materials, specimens and cultures must be decontaminated before disposal or cleaning for reuse. They should be placed in leak-proof, colour-coded plastic bags for autoclaving or incineration on the premises. These bags should be supported in rigid containers. If it is necessary to move the bags to another site for decontamination they should be placed in leak-proof (i.e., solid-bottomed) containers which can be closed before they are removed from the laboratory.

10. Laboratory coveralls, gowns or uniforms must be worn for work in the laboratory. This clothing should not be worn in non-laboratory areas such as offices, libraries, staff rooms and canteens. Contaminated clothing must be decontaminated by appropriate methods.

11. Open-toed footwear should not be worn.

12. Protective laboratory clothing should not be stored in the same lockers or cupboards as street clothing.

13. Safety glasses, face shields (visors) or other protective devices must be worn when it is necessary to protect the eyes and face from splashes and impacting objects.

14. Only persons who have been advised of the potential hazards and who meet specific entry requirements (e.g., immunization) should be allowed to enter the laboratory working areas. Laboratory doors should be kept closed when work is in progress; access to animal houses should be restricted to authorized persons; children should be excluded from laboratory working areas.

15. There should be an arthropod and rodent control programme.

16. Animals not involved in the work of the laboratory should not be permitted in or near the laboratory.

17. The use of hypodermic needles and syringes should be restricted to parenteral injection and aspiration of fluids from laboratory animals. Their use for removing the contents of diaphragm bottles should be limited (opening devices are available, permitting the use of pipettes). Hypodermic needles and syringes should not be used as substitutes for pipetting devices in the manipulation of infectious fluids. Cannulas should be used instead of needles whenever possible.

18. Gloves appropriate for the work must be worn for all procedures that may involve accidental direct contact with blood, infectious materials or infected animals. After use, gloves should be removed aseptically and autoclaved with other laboratory wastes before disposal. Hands must then be washed. Reusable gloves must be washed while on the hands and after removal, and cleaned and disinfected before reuse (see Chapter 9, page 60).

19. All spills, accidents and overt or potential exposures to infectious materials must be reported immediately to the laboratory supervisor. A written record of such accidents and incidents should be maintained.

20. Appropriate medical evaluation, surveillance and treatment should be provided.

21. Baseline serum samples may be collected from laboratory staff and other persons at risk. These should be stored as appropriate. Additional specimens should be collected periodically depending on the organisms handled and the function of the laboratory.

22. The laboratory supervisor should ensure that training in laboratory safety is provided. A safety or operations manual that identifies known and potential hazards and that specifies practices and procedures to minimize or eliminate such hazards should be adopted. Personnel should be advised of special hazards and required to read and follow standard practices and procedures. The supervisor should make sure that personnel understand these.

Laboratory design and facilities

In designing a laboratory and assigning certain types of work to it, special attention should be paid to conditions that are known to pose problems. These include:

— formation of aerosols;
— work with large volumes and/or high concentrations of microorganisms;
— overcrowding and overequipping;
— infestation with rodents and arthropods;
— unauthorized entrance.

Design features

1. Ample space must be provided for the safe conduct of laboratory work and for cleaning and maintenance.

2. Walls, ceilings and floors should be smooth, easily cleanable, impermeable to liquids, and resistant to the chemicals and disinfectants normally used in the laboratory. Floors should be slip-resistant. Exposed pipes and ducting should stand clear of the walls. Horizontal surfaces should be avoided where possible to prevent accumulation of dust.

3. Illumination should be adequate for all activities. Undesirable reflections and glare should be avoided.
4. Bench tops should be sealed to the walls, impervious to water and resistant to disinfectants, acids, alkalis, organic solvents and moderate heat.
5. Laboratory furniture should be sturdy. Open spaces between and under benches, cabinets and equipment should be accessible for cleaning.
6. Storage space must be adequate to hold supplies for immediate use and thus prevent clutter on bench tops and in aisles. Additional long-term storage space, conveniently located outside the working areas, should also be provided.
7. Hand-wash basins, with running water if possible, should be provided in each laboratory room, preferably near the door.
8. Doors should have appropriate fire ratings, be self-closing, and have vision panels.
9. An autoclave (or a suitable pressure cooker) should be available in the same building as the laboratory.
10. Facilities for storing outer garments and personal items and for eating and drinking should be provided outside the working areas.
11. There are no specific ventilation requirements. In the planning of new facilities, however, consideration should be given to the provision of mechanical ventilation systems that provide an inward flow of air without recirculation. If there is no mechanical ventilation, windows should be openable and preferably fitted with arthropod-proof screens. Skylights should be avoided.
12. Space and facilities should be provided for the safe handling and storage of solvents, radioactive materials, and compressed and lique-fied gases.
13. Safety systems should cover fire, electrical emergencies, emergency shower, and eyewash facilities.
14. First-aid areas or rooms suitably equipped and readily accessible should be available.
15. A good-quality and dependable water supply is essential. There should be no cross-connections between sources of laboratory and drinking-water supplies. The public water system should be protected by an anti-backflow device.
16. There should be a reliable and adequate electricity supply and emergency lighting to permit safe exit. A stand-by generator is desirable for the support of essential equipment such as incubators, biological safety cabinets, freezers, etc., and for the ventilation, where necessary, of animal cages.

17. There should be a reliable and adequate supply of town, natural or bottled gas. Good maintenance of the installation is mandatory.
18. Three aspects of waste disposal need special attention to meet performance and pollution-control requirements:

— autoclaves for the treatment of solid wastes need specially designed accommodation and services;
— incinerators should be of special design, equipped with afterburners and smoke-consuming devices;
— it may be necessary to decontaminate wastewater.

19. Laboratories and animal houses are occasionally the targets of vandals. Fire and physical security should be considered to avoid conflict. Security may be augmented by strong doors, screened windows and restricted issue of keys.

Further information about laboratory design and facilities is given elsewhere (*1–5*).

Laboratory equipment

The risk of infection can be minimized by the use of safety equipment, facilities and practices. This section deals primarily with laboratory equipment suitable for work with Risk Group 2 microorganisms (and also those in Risk Group 3: see pages 19–23).

The head of the laboratory should, after consultation with the safety officer and safety committee (see pages 99–101), ensure that adequate equipment is provided and that it is used properly. Equipment should be selected to take account of certain general principles, i.e., it should be

— designed to prevent or limit contact between the operator and the infectious material;
— constructed of materials that are impermeable to liquids, resistant to corrosion, and meet structural requirements;
— fabricated to be free of burrs, sharp edges and unguarded moving parts;
— designed, constructed and installed to facilitate simple operation and provide for ease of maintenance, cleaning, decontamination and certification testing.

Detailed performance and construction specifications may also be required to ensure that the equipment will possess the necessary safety features.

Recommended biosafety equipment

1. Pipetting aids—to replace mouth pipetting. Many different designs are available (see page 82).
2. Biological safety cabinets (see page 78), to be used whenever:

   — Procedures with a high potential for producing aerosols are used and there is a risk of airborne infection. These may include centrifugation, grinding, blending, vigorous shaking or mixing, sonic disruption, opening of containers of infectious materials whose internal pressure may be different from the ambient pressure, intranasal inoculation of animals, and harvesting of infected tissues from animals and eggs.
   — High concentrations or large volumes of infectious materials are handled. Such materials may be centrifuged in the open laboratory if sealed centrifuge safety cups (pages 39–40) are used and if they are loaded and unloaded in a biological safety cabinet.

3. Transfer loop incinerators—to reduce aerosol production. Plastic disposable transfer loops are now available.
4. Screw-capped tubes and bottles—to hold positive specimens and cultures.
5. Autoclaves—to sterilize contaminated material (pages 65–67).
6. Where available, plastic, not glass, Pasteur pipettes.

Health and medical surveillance

The objectives of the health and medical surveillance of laboratory personnel are:

— to provide a means of preventing occupationally acquired disease in healthy people by the exclusion of highly susceptible individuals as well as by regular review of those accepted for employment;
— to provide active or passive immunization where indicated;
— to provide a means for the early detection of laboratory-acquired infections;
— to assess the efficacy of protective equipment and procedures.

It is the responsibility of the employing authority through the laboratory director to ensure that there is adequate surveillance of the health of laboratory personnel.

Guidelines for the surveillance of workers handling microorganisms in Risk Group 1

These microorganisms are unlikely to cause human disease or animal disease of veterinary importance. Ideally, however, there should be a pre-employment health check which should include the person's medical history. Prompt reporting of illnesses or laboratory accidents is desirable and all staff members should be made aware of the importance of maintaining good microbiological techniques.

Guidelines for the surveillance of workers handling microorganisms in Risk Group 2

1. A pre-employment or preplacement health check is necessary. This screening should include medical history. A clinical examination and the collection of a baseline serum sample would be useful and, in some cases, may be necessary.
2. The laboratory should maintain an up-to-date list of the employees' family medical practitioners.
3. Records of illness and absence should be kept by the laboratory director, and it is the responsibility of the laboratory worker and his or her own medical adviser to keep the director informed of all absences due to illness.
4. Women of childbearing age should be made aware, in unequivocal terms, of the risk to the unborn child of occupational exposure to certain microorganisms, e.g., rubella virus. The precise steps taken to protect the fetus will vary, depending on the microorganisms to which the women may be exposed.

Further information about medical surveillance of laboratory personnel is given elsewhere *(6, 7, 8)*.

Training

Human error and poor technique can compromise the best of safeguards and equipment provided specifically to protect the laboratory worker. Thus, a safety-conscious staff, well informed about the recognition and control of the hazards present in the laboratory, is a key element in the prevention of laboratory accidents and acquired infections. For this reason, continuous in-service training in safety measures is essential. The process begins with the laboratory management, which should ensure that safe laboratory practices and procedures are integrated into the basic training of employees. Safety measures should always be an integral part of a new employee's introduction to the laboratory.

Laboratory supervisors must play the key role in training their immediate staff in good laboratory techniques. The safety officer can assist in training and with the development of training aids and documentation.

Staff training should always include safe methods of dealing with the following hazardous procedures that are commonly encountered by all laboratory personnel:

— procedures involving inhalation risks (i.e., aerosol production), such as using loops, streaking agar plates, pipetting, making smears, opening cultures, centrifugation;
— procedures involving ingestion risks, such as handling specimens, smears and cultures;
— procedures involving injection risks, such as syringe and needle techniques, and animal handling that may result in bites and scratches;
— procedures for the safe handling of blood and other potentially hazardous pathological materials;
— procedures involving the disposal of infectious material.

For additional information see pages 104–110.

Decontamination and disposal

In laboratories, decontamination and disposal are closely interrelated. All materials will ultimately be disposed of but, in terms of daily use, only some of them will require actual removal from the laboratory or destruction. The remainder, e.g., some glassware, instruments and laboratory clothing, will be recycled and reused. Disposal should therefore be interpreted in the broad sense, rather than the restrictive sense of a destructive process.

The principal questions to be asked before disposal of any object or materials from laboratories that deal with potentially infectious micro-organisms or animal tissues are:

— Have the objects or materials been effectively sterilized or disinfected by an approved procedure?
— If not, have they been packaged in an approved manner for immediate on-site incineration or transfer to another laboratory?
— Does the disposal of the sterilized or disinfected objects or materials involve any additional potential hazards, biological or otherwise, to those who carry out the immediate disposal procedures or those who might come into contact with them outside the laboratory complex?

15

Decontamination

Materials for decontamination and disposal should be placed in containers, e.g. autoclavable plastic bags that are colour-coded according to whether the contents are to be autoclaved or incinerated.

Autoclaving is the preferred method for all decontamination processes. The autoclave should be of the gravity displacement type.

Alternative methods, if an autoclave is not available, include:

— use of a pressure cooker at the highest attainable working pressure;
— boiling for 30 minutes, preferably in water containing sodium bicarbonate.

For additional information see pages 60–70.

Disinfectants and chemicals

There should be a written policy stating which disinfectants are to be used for what purposes, and the manufacturers' recommended dilution for each.

Sodium hypochlorite and phenolic compounds are the disinfectants recommended for general laboratory use (see Table 6, p. 64).

For special purposes, various surface-active or lipid-destroying agents, including alcohols, iodine, iodophors and other oxidizing agents, as well as a very high or extremely low pH, can be effective provided that it has been established that the agent to be destroyed is not resistant to the procedure.

Other methods

The use of dry heat is discouraged because of its unpredictable variation. Similarly, microwave, ultraviolet and ionizing radiation are unsuitable. For further information and references see pages 60–70.

Disposal

An identification and separation system for contaminated materials (and their containers) should be established. Categories may be:

(a) non-contaminated waste that can be disposed of with general waste;
(b) "sharps"—hypodermic needles, scalpels, knives, broken glass;
(c) contaminated material for autoclaving and recycling;
(d) contaminated material for disposal;
(e) anatomical waste, e.g., human and animal tissue.

Sharps

Hypodermic needles should not be recapped, clipped, or removed from disposable syringes. The assembly should be placed in a container with impenetrable walls. These containers must not be filled to capacity. When they are three-quarters full they should be placed in a "contaminated waste" container and incinerated, with prior autoclaving if laboratory practice requires it.

Disposable syringes, used alone, should be placed in containers and incinerated, with prior autoclaving if required.

Contaminated materials for autoclaving and recycling

Note. No precleaning should be attempted—any necessary cleaning or repair should be done after autoclaving.

Contaminated materials for disposal

All cultures and contaminated materials should normally be autoclaved in leakproof containers, e.g., autoclavable, colour-coded plastic bags, before disposal. After autoclaving, the material may be placed in transfer containers for transport to the incinerator or other point of disposal.

Discard pots, pans or jars, preferably unbreakable (e.g., plastic), and containing a suitable disinfectant, freshly prepared each day, should be placed at every work station. Waste materials should remain in intimate contact with the disinfectant (i.e., not protected by air bubbles) for at least 18 hours. The disinfectant may then be poured down a sluice or sink and the solid contents placed in a container for autoclaving or incineration. The discard pots should be autoclaved and washed before reuse.

If an incinerator is available on the laboratory site, autoclaving may be omitted. The contaminated waste should be placed in designated containers (e.g., colour-coded bags) and transported directly to the incinerator. If plastic bags are used they should be placed in cardboard boxes for transport. Reusable transfer containers should be leakproof and have tight-fitting covers. They should be disinfected and cleaned before they are returned to the laboratory for further use.

Incineration is the method of choice for the final disposal of contaminated waste, including carcasses of laboratory animals, preferably after autoclaving (see pages 69–70). Incineration of contaminated waste must meet with the approval of the public health and air pollution authorities as well as that of the safety officer.

When incinerators are not approved for such use, final disposal methods must be established in cooperation with the authorities concerned.

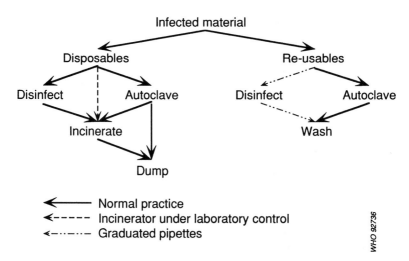

Fig. 2. Flow chart for the treatment of infected material
Source: ref *5*. Reproduced by kind permission of the publisher.

Fig. 2 is a flow chart for the treatment of contaminated materials. More detailed information about disinfection and sterilization and references can be found in Chapter 9, page 60.

Chemical, fire, electrical and radiation safety

A breakdown in the containment of pathogenic organisms may be the indirect result of chemical, fire, electrical or radiation accidents. It is therefore essential to maintain high standards of safety in these fields in any microbiological laboratory.

Statutory rules and regulations for each of these will normally be laid down by the competent national or local authority, whose assistance should be sought if necessary. Chemical, fire and electrical hazards are considered in Part 4 of this manual. A preliminary assessment of the status of a laboratory in respect of these hazards may be made by applying the safety checklist given in Part 6.

3. The Containment laboratory – Biosafety Level 3

The Containment laboratory – Biosafety Level 3 is designed and provided for work with Risk Group 3 microorganisms, i.e. those that present a high risk to laboratory workers but a low risk to the community.

This level of containment requires strengthening of the operational and safety programmes of Basic laboratories – Biosafety Levels 1 and 2, as well as the provision of structural safeguards and the mandatory use of biological safety cabinets.

These guidelines are presented in the form of additions to those for Basic laboratories, which must therefore be applied before those specific for the Containment laboratory – Biosafety Level 3. The major additions and changes are in:

— code of practice;
— laboratory design and facilities;
— health and medical surveillance.

Laboratories in this category should be registered or listed with the national or other appropriate health authorities.

Code of practice

The code of practice for Basic laboratories – Biosafety Levels 1 and 2 applies except where modified as follows:

1. The two-person rule should apply, whereby no individual ever works alone in the laboratory.
2. The biohazard warning sign (see Fig. 1) displayed on laboratory doors must identify the microorganisms handled and the name of the laboratory supervisor who controls access, and indicate any special conditions of entry into the area, e.g., immunization.
3. Laboratory clothing that protects street clothing, i.e., solid-front or wrap-around gowns, scrub suits, coveralls, head covering, and where

appropriate, shoe covers or dedicated shoes, must be worn in the laboratory. Front-buttoned laboratory coats are unsuitable. Laboratory clothing must not be worn outside the laboratory and it must be decontaminated before it is laundered.

4. When appropriate, respiratory equipment must be worn in rooms containing infected animals.

Laboratory design and facilities

The Containment laboratory – Biosafety Level 3 is designed for work with Risk Group 3 microorganisms and with large volumes and high concentrations of Risk Group 2 microorganisms, where there is a high risk of aerosol spread or infection.

The section on design and facilities for Basic laboratories – Biosafety Levels 1 and 2 (see page 10–12) applies except where modified as follows.

1. The laboratory should be separated from the areas that are open to unrestricted traffic flow within the building. Additional separation may be achieved by placing the laboratory at the blind end of a corridor, or constructing a partition and door or access through an anteroom or Basic laboratory – Biosafety Level 2.

2. Entry for personnel must be through a vestibule (i.e., double-door entry).

3. Access to the laboratory area must be designed to prevent entrance of arthropods and other vermin.

4. Access doors must be self-closing and lockable. A break-through panel may be provided for emergency use.

5. The surfaces of walls, floors and ceilings should be water-resistant and easy to clean. Openings in these surfaces (e.g., for service pipes) should be sealed to facilitate decontamination of the room(s).

6. The laboratory room must be sealable for decontamination. Air-ducting systems must be constructed to permit gaseous disinfection.

7. Windows must be closed and sealed.

8. A foot- or elbow-operated hand-wash basin should be provided near to each exit door.

9. There must be a ventilation system that establishes a directional air flow from access spaces into the laboratory room. Staff must verify that proper directional air flow into the laboratory room is achieved.

10. In laboratories that have supply air systems, the supply and exhaust facilities must be interlocked to ensure inward air flow at all times.

11. The building ventilation system must be so constructed that air from the Containment laboratory – Biosafety Level 3 is not recirculated to other areas within the building. Air may be reconditioned and recirculated within that laboratory. Exhaust air from the laboratory

(other than from biological safety cabinets) must be discharged directly to the outside of the building so that it is dispersed away from occupied buildings and air intakes. It is recommended that this air is discharged through high-efficiency particulate air (HEPA) filters (see page 82).

12. Biological safety cabinets should be sited away from walking areas and out of cross-currents from doors, windows and ventilation systems (see page 78).

13. The exhaust air from Class I or Class II biological safety cabinets, which will have been passed through HEPA filters, must be discharged directly or through the building system to the outside air. If it is discharged through the building system it must be connected to this system in such a way as to avoid interference with the air balance of the cabinet or the building exhaust system. HEPA filters must be installed in a manner that permits gaseous decontamination and aerosol test challenge.

14. An autoclave for the decontamination of infected waste material should be available in the laboratory room. If such wastes have to be transported to an autoclave in another part of the same building for decontamination, they should be held in a covered, leakproof container.

15. Anti-backflow devices must be fitted to the water supply.

16. Liquid effluents must be discharged directly to the sanitary sewer.

For further information see references *1, 2, 3, 5, 9,* and *10.*

Laboratory equipment

The principles for the selection of equipment, including biological safety cabinets (see page 78), are the same as for the Basic laboratory – Biosafety Level 2 (page 12) except that all activities involving infectious materials are conducted in biological safety cabinets, with other physical containment devices, or special personal protective equipment.

While Class I or Class II biological safety cabinets are normally used in the Containment laboratory – Biosafety Level 3, a Class III biological safety cabinet may be needed for some procedures with Risk Group 3 microorganisms.

Health and medical surveillance

The objectives of health and medical surveillance programmes for Basic laboratories – Biosafety Levels 1 and 2 (see page 13) apply to Containment laboratories – Biosafety Level 3, except where modified as follows.

A. Front of card

ILLNESS SURVEILLANCE NOTICE

_____

Name

_____

TO THE EMPLOYEE
Keep this card in your possession. In case of unexplained
febrile illness, present the card to your physician and notify one
of the following in the order listed.

_____

	Work
Dr	Home
	Work
Dr	Home

B. Back of card

TO THE PHYSICIAN
The holder of this card works in an area at ———————
—— in which pathogenic viruses, rickettsia, or bacteria are
present. In the event of an unexplained febrile illness, please
call the employer for information on agents to which this
employee may have been exposed.

Name of laboratory: ———————————————

Address: ——————————————————————

———————————————————

———————————————————

Tel: ——————————————————————

Fig. 3. Suggested format for medical contact card

1. Medical examination of all laboratory personnel who work in the Containment laboratory – Biosafety Level 3 is mandatory. This should include a detailed medical history and a physical examination.
2. A baseline serum sample should be obtained and stored for future reference.
3. Employees who are immunocompromised should not be employed in Containment laboratories – Biosafety Level 3.
4. Special consideration should be given to the employment of pregnant women (see page 14).

After a satisfactory clinical assessment report, the examinee should be provided with a medical contact card (Fig. 3) stating that he or she is employed in a Containment laboratory – Biosafety Level 3. It is suggested that this card be wallet-sized and that it should always be carried by the holder.

Note. The name(s) of the contact persons to be entered will need to be agreed locally but might include the laboratory director, medical adviser or biosafety officer.

For further information on medical surveillance see references *6, 7,* and *8.*

4. The Maximum Containment laboratory – Biosafety Level 4

The Maximum Containment laboratory – Biosafety Level 4 is designed for work with infectious materials or experiments in microbiology that present, or are suspected of presenting, a high risk to both the laboratory worker and the community.

Before such a laboratory is constructed and put into operation, intensive consultations should be held with institutions that have had experience of operating a similar facility. Operational Maximum Containment laboratories – Biosafety Level 4 should be under the control of national or other appropriate health authorities.

Laboratory design and facilities

The features of a Containment laboratory – Biosafety Level 3 apply to a Maximum Containment laboratory – Biosafety Level 4 with the addition of the following (1, 2):

1. *Controlled access.* Entry and exit of personnel and supplies must be through an airlock or pass-through system. On entering, personnel should put on a complete change of clothing; before leaving, they should shower before putting on their street clothing.
2. *Controlled air system.* Negative pressure must be maintained in the facility by a mechanical, individual, inwardly directed, HEPA-filtered supply, and an exhaust air system with HEPA filters in the exhaust and, where necessary, in the intake.
3. *Decontamination of effluents.* All fluid effluents from the facility, including shower water, must be rendered safe before final discharge.
4. *Sterilization of waste and materials.* A double-door, pass-through autoclave must be available.
5. *Primary containment.* An efficient primary containment system must be in place, consisting of one or more of the following: (*a*) Class III biological safety cabinets, (*b*) positive-pressure ventilated suits. In the

latter case a special chemical decontamination shower must be provided for personnel leaving the suit area.

6. *Airlock entry ports for specimens and materials.*

Fig. 4 shows the essential elements of a Maximum Containment laboratory – Biosafety Level 4.

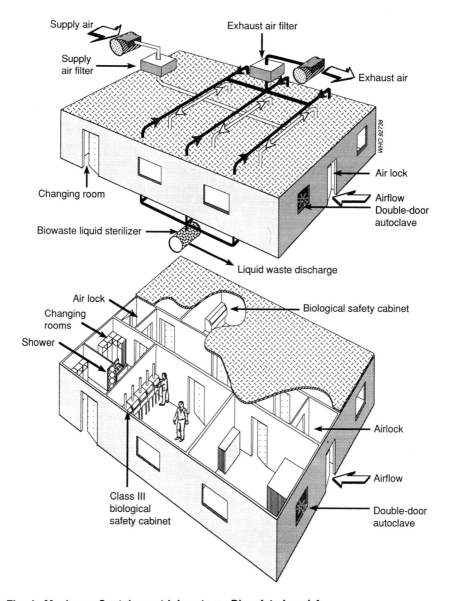

Fig. 4. Maximum Containment laboratory – Biosafety Level 4

Because of the great complexity of the work, a detailed work manual should be developed and tested in training exercises. In addition, an emergency programme must be devised (see also Chapter 8, page 55). In the preparation of this programme, active cooperation with national and local health authorities should be established. Other emergency services, e.g., fire, police, receiving hospitals, should also be involved.

5. Laboratory animal facilities

Those who use animals for experimental and diagnostic purposes have a moral obligation to take every care to avoid causing them unnecessary pain or suffering. The animals must be provided with comfortable, hygienic housing and adequate, wholesome food and water. At the end of the experiment they must be killed in a humane and painless manner.

The animal house should be an independent, detached unit. If it adjoins a laboratory, the design should provide for its isolation from the public parts of the laboratory should such need arise, and for its decontamination and disinfestation.

Table 4. Animal facility containment levels; summary of practices and safety equipment

Risk group	Biosafety level	Laboratory practices	Safety equipment
1	1	Limited access, protective clothing and gloves	
2	2	Limited access and hazard warning signs; protective clothing and gloves; decontamination of wastes and cages before washing	Class I or II BSC[a] for activities that produce aerosols; personal protection devices
3	3	Controlled access; special protective clothing; otherwise as for levels 1 and 2	Class I and II BSCs for all activities; personal protection devices
4	4	Strictly limited access; level 3 practices plus clothes changing room and shower; all wastes decontaminated before removal from facility	Class III BSCs or positive-pressure suits for all activities

[a] BSC, biological safety cabinet.

Animal facilities, like laboratories, may be designated primarily according to the risk group of the agents under investigation as Biosafety Level 1, 2, 3, or 4 (*1, 2, 11*). Other factors, however, should also be taken into consideration. In respect of the agents, these include the volumes and concentrations to be used, the route of inoculation and whether and by what route they may be excreted. In respect of the animals, they include the nature of the animals, i.e., their aggressiveness and tendency to bite and scratch, their natural ecto- and endoparasites, the zoonotic diseases to which they are susceptible, and the possible dissemination of allergens.

As with laboratories, the requirements for design features, equipment and precautions increase in stringency according to the biosafety level. These are described below and summarized in Table 4.

Animal facility – Biosafety Level 1

This is suitable for the maintenance of most stock animals after quarantine (except nonhuman primates) and for animals that are deliberately inoculated with agents in Risk Group 1. Good microbiological technique (GMT) is required.

Animal facility – Biosafety Level 2

This is suitable for work with animals that are deliberately inoculated with agents in Risk Group 2. The following safety precautions apply.

1. Biohazard warning signs should be posted on doors and other appropriate places.
2. The facility must be designed for easy cleaning and housekeeping.
3. Doors must open inwards and be self-closing.
4. Heating, ventilation and lighting must be adequate.
5. If mechanical ventilation is provided, the air flow must be inwards, maintained by extracting air to the atmosphere. Air should *not* be recirculated to any part of the building, i.e., "total loss" system.
6. Access must be restricted to authorized persons.
7. No animals should be admitted other than those for experimental use.
8. There should be an arthropod and rodent control programme.
9. Arthropod screens must be fitted.
10. Work surfaces must be decontaminated after use with effective disinfectants.
11. Biological safety cabinets (Classes I or II) must be provided for work that may involve the generation of aerosols.
12. An autoclave must be available on site or nearby.
13. Animal bedding materials must be removed in a manner that minimizes the generation of aerosols and dust.

14. All waste materials and bedding must be decontaminated before disposal.
15. Material for autoclaving or incineration must be transported safely in closed containers.
16. Animal cages must be decontaminated after use.
17. Animal carcasses must be incinerated.
18. Hand-washing facilities must be provided. Staff must wash their hands before leaving the animal house.
19. Protective clothing must be worn in the facility, and discarded on leaving. Suitable gloves should be available.
20. All injuries, however minor, must be reported and recorded.
21. Eating, drinking, smoking and application of cosmetics must be forbidden in the facility.

Animal facility – Biosafety Level 3

This is suitable for work with animals that are deliberately inoculated with agents in Risk Group 3.

1. All the requirements for animal facilities – Biosafety Levels 1 and 2 must be implemented.
2. Access must be strictly controlled.
3. The facility must be separated from other laboratory and animal house areas by an anteroom with two doors, forming an airlock.
4. Hand-washing facilities and showers must be provided in the anteroom.
5. There must be mechanical ventilation to ensure a continuous air flow through all the rooms. Exhaust air must pass through HEPA filters before being discharged to the atmosphere (total loss). The system must be designed to prevent accidental reverse flow and positive pressurization in any part of the animal house.
6. An autoclave must be provided within the facility.
7. An incinerator should be readily available on site or alternative arrangements should be made with the authorities concerned (see pages 69–70).
8. Animals infected with Risk Group 3 agents must be housed in cages in isolators or rooms with ventilation exhausts placed behind the cages.
9. Bedding should be as dust-free as possible.
10. Single-wear protective clothing should be worn in the facility. It should be discarded on leaving the facility, and autoclaved before disposal.
11. Immunization of staff, as appropriate, should be considered.

Animal facility – Biosafety Level 4

Work in this facility will normally be linked with that in the Maximum Containment laboratory – Biosafety Level 4 and national and local rules should be harmonized to apply to both.

1. All the requirements for animal facilities – Biosafety Levels 1, 2 and 3 must be implemented as appropriate.
2. Access must be strictly controlled (key entry); only staff designated by the director of the establishment should have authority to enter.
3. Individuals must not work alone: the two-person rule must apply.
4. Personnel must have received the highest possible level of training as microbiologists and be familiar with the hazards involved in their work and the necessary precautions.
5. If the facility is not part of a Maximum Containment labora-tory – Biosafety Level 4 it must be an isolated building.
6. The facility must be entered by an airlock antechamber, the clean side of which must be separated from the restricted side by changing and showering facilities.
7. The facility must be ventilated by a HEPA-filtered exhaust system designed to ensure a negative pressure.
8. The ventilation system must be designed to prevent reverse flow and over-pressurization.
9. A double-ended autoclave must be provided with the clean end in a room outside the facility rooms.
10. Staff must remove street clothing when entering and put on special, single-wear, protective clothing. After work they must discard the protective clothing to a bin for autoclaving and disposal, and shower before leaving.
11. A pass-through airlock must be provided for delivery of materials.
12. All manipulations with animals must be done in Class III biological safety cabinets.
13. All animals must be housed in isolators.
14. All bedding and waste must be autoclaved before leaving the facility.
15. There must be medical supervision of staff and immunization as appropriate.

Further information about the design and equipment of animal house facilities is given elsewhere (*1, 2, 7, 11–15*).

Invertebrates

The invertebrates that are used for experimental purposes in laboratories are usually the reservoirs or vectors of pathogens or, as in the case of ecological and environmental investigations, may be fortuitously infected with pathogens ingested with their food.

The invertebrates used in laboratories include: Annelida, Aschelminthes, Arthropoda, Echinodermata, Mollusca, Platyhelminthes and Protozoa.

As with vertebrates, the animal facility biosafety level will normally be determined by the risk group of the agent under investigation or naturally present, but some special and additional precautions are necessary with certain arthropods, particularly with flying insects, as follows.

1. Separate rooms should be provided for infected and noninfected invertebrates.
2. The rooms should be capable of being sealed for fumigation.
3. Insecticide sprays should be readily available.
4. "Chilling" facilities should be provided to reduce, where necessary, the activity of invertebrates.
5. Access should be through an anteroom with arthropod-proof screens on the doors and insect traps.
6. All exhaust ventilation ducts and openable windows should be fitted with arthropod-proof screens.
7. Waste traps on sinks and sluices should not be allowed to dry out.
8. All waste should be decontaminated by heat, as some invertebrates are not killed by all disinfectants.
9. A check should be kept on the numbers of larval and adult forms of flying, crawling and jumping arthropods.
10. Containers for ticks and mites should stand in trays of oil.
11. Flying insects infected with agents in higher risk groups must be contained in double-netted cages.
12. Arthropods infected with agents in higher risk groups must be handled in biological safety cabinets or isolators.

For further information on facilities for housing invertebrates see references *2* and *16*.

PART 2

Good microbiological technique

6. Safe laboratory techniques

Human error, poor techniques and misuse of equipment cause the majority of laboratory accidents and related infections. This section provides a compendium of technical methods that are designed to avoid or minimize the most commonly reported accidents caused by these factors. The reader is also referred to the sections on codes of practice, equipment and training. Other information is available elsewhere (*5, 9, 10, 16, 17–20*).

Techniques for the safe handling of specimens in the laboratory

Improper collection, internal transport and receipt of specimens in the laboratory carry a risk of infection to the personnel involved.

Specimen containers

Specimen containers may be of glass or plastic. They should be robust and should not leak when the cap or stopper is correctly applied. No material should remain on the outside of the container.

Containers should be correctly labelled to facilitate identification in the laboratory. Ideally, the containers should be placed in plastic bags, but such bags should not be stapled.

Specimen request forms should not be wrapped around the containers but placed in separate, preferably plastic envelopes.

Some institutions require specimens suspected of containing pathogens in Risk Group 3, hepatitis B virus or human immunodeficiency virus to be identified on the container and request form by a special "danger of infection" label.

Transport to the laboratory

To avoid accidental leakage or spillage, special secondary containers, such as trays or boxes, should be used, fitted with racks so that the specimen containers remain upright. The secondary containers may be of metal or plastic but should be autoclavable or resistant to the action of chemical disinfectants. They should be regularly decontaminated.

Receipt of specimens

Laboratories that receive large numbers of specimens should designate a particular room or area for their receipt.

Opening packages

Personnel who receive and unpack specimens should be aware of the potential health hazards involved, and should seek professional assistance for dealing with broken or leaking containers. Disinfectants should be available.

Specimens should be unpacked on trays. Any specimen carrying a "danger of infection" label or received by air freight or air mail should ideally be unpacked in a biological safety cabinet.

Techniques for the use of pipettes and pipetting aids

1. A pipetting aid should always be used. Pipetting by mouth should be prohibited.
2. All pipettes should have cotton plugs to reduce contamination of pipetting devices.
3. Air should never be blown through a liquid containing infectious agents.
4. Infectious materials should not be mixed by alternate suction and expulsion through a pipette.
5. Liquids should not be forcibly expelled from pipettes.
6. To avoid dispersion of infectious material if it is accidentally dropped from a pipette, a disinfectant-soaked cloth or absorbent paper should be placed on the working surface; this should be autoclaved after use.
7. Mark-to-mark pipettes are preferable to other types, as they do not require expulsion of the last drop.
8. Contaminated pipettes should be completely submerged in a suitable disinfectant (pages 60–63) contained in an unbreakable container. They should be left for 18–24 hours before disposal.
9. A discard container for pipettes should be placed within the biological safety cabinet, not outside it.

10. Syringes fitted with sharp hypodermic needles must not be used for pipetting. Blunt cannulas should be used instead of needles. There are devices for opening septum-capped bottles which avoid the use of hypodermic needles and syringes.

Techniques for avoiding the dispersal of infectious materials

1. The circle of a microbiological transfer loop should be completely closed and the shank should be not more than 6 cm in length.
2. The risk of spatter of infectious material in a Bunsen flame should be avoided by using a microincinerator for flaming transfer loops. Disposable transfer loops, which do not need flaming, are preferable.
3. Catalase tests should not be done on slides. Tube or cover-glass methods should be used. Another method is to touch the surface of a colony with a microhaematocrit tube loaded with hydrogen peroxide.
4. Discarded specimens and cultures for disposal should be placed in leakproof containers, e.g., laboratory discard bags, suitably supported.
5. Working areas must be decontaminated with a suitable disinfectant at the end of each work period.

Techniques for the use of biological safety cabinets

1. The use and limitations of biological safety cabinets (page 78) should be explained to all potential users, with reference to national standards and relevant literature. Written protocols should be issued to staff. In particular, it must be made clear that the cabinet will not protect the hands of the worker from gross spillage, breakage, or poor technique.
2. The cabinet must not be used unless it is working properly.
3. The glass viewing panel must not be opened when the cabinet is in use.
4. Apparatus and materials in the cabinet must be kept to a minimum and at the rear of the working area.
5. A Bunsen burner must not be used in the cabinet. The heat produced will distort the air flow and may damage the filters. A microincinerator is permissible but disposable transfer loops are better.
6. All work must be done in the middle or rear part of the working surface and be visible through the viewing panel.
7. Traffic behind the operator should be minimized.
8. The operator should not disturb the air flow by repeated removal and reintroduction of his or her arms.
9. The cabinet fan should be run for at least 5 minutes after completion of work in the cabinet.

Note. Horizontal outflow cabinets ("clean-air work stations") are **not** biological safety cabinets and should not be used as such.

Further information about biological safety cabinets is given in Chapter 11. See also references *1, 5, 11* and *21–27*.

Techniques for avoiding ingestion of infectious materials and their contact with skin and eyes

1. Large particles and droplets ($> 5 \mu m$) released during microbiological manipulations settle rapidly on bench surfaces and on the hands of the operator. Hands should be washed frequently. Workers should avoid touching their mouths and eyes.
2. Food and drink must not be consumed or stored in the laboratory.
3. There should be no smoking or gum-chewing in the laboratory.
4. Cosmetics should not be applied in the laboratory.
5. The face and eyes should be shielded or otherwise protected during any operation that might result in the splashing of infectious materials.

Techniques for avoiding injection of infectious materials

1. Injection may result from accidents with hypodermic needles ("needle-stick"), glass Pasteur pipettes and broken glass.
2. Needle-stick accidents can be reduced only by (*a*) taking particular care, and (*b*) minimizing the use of syringes and needles; for many techniques, syringes with blunt cannulas may be used instead. Simple devices are available for opening septum-stoppered bottles so that pipettes can be used.
3. Soft plastic Pasteur pipettes should replace those made of glass.
4. Accidental inoculation with broken or chipped glassware can be avoided only by personal care.

Techniques for the separation of serum

1. Only properly instructed staff should be employed for this work.
2. Gloves should be worn.
3. Splashes and aerosols can be avoided or minimized only by good technique. Blood and serum should be pipetted carefully, not poured. Pipetting by mouth should be forbidden.
4. Pipettes should be completely submerged in hypochlorite or other suitable disinfectant. They should remain in the disinfectant at least overnight before disposal or decontamination for reuse.

5. Discarded specimen tubes containing blood clots, etc. (with their caps replaced), should be placed in suitable leakproof containers for autoclaving and incineration.

6. A solution of hypochlorite, freshly prepared daily, should be available for dealing with splashes and spillages of blood and serum (see Chapter 9).

Techniques for the use of centrifuges

General

1. Satisfactory mechanical performance is a prerequisite of microbiological safety in the use of laboratory centrifuges.

2. Centrifuges should be operated according to the manufacturers' instructions.

3. Centrifuges should be placed at such a level that workers of less than average height can see into the bowl to place trunnions and buckets correctly.

4. Centrifuge rotors and buckets should be inspected daily for signs of corrosion and for hair-line cracks.

5. Buckets and trunnions should be paired by weight and, with tubes in place, correctly balanced.

6. Alcohol (propanol, 70%) should be used for balancing buckets. Saline or hypochlorite solution should not be used as both corrode metals.

7. After use, buckets should be stored in an inverted position to drain the balancing fluid.

8. Infectious airborne particles may be ejected when centrifuges are used. These particles travel at speeds too high to be retained if the centrifuge is placed in a traditional Class I or Class II biological safety cabinet.

9. Good centrifuge technique, securely capped tubes and sealable centrifuge buckets ("safety cups") offer adequate protection against infectious aerosols and dispersed particles of microorganisms in Risk Groups 2, 3 and 4.

For further information about safe work with centrifuges, see references *28* and *29*.

Centrifugation of Risk Group 2 microorganisms and materials

1. Centrifuge tubes and specimen containers for use in the centrifuge should be made of thick-walled glass or of plastic and should be inspected for defects before use.

2. Tubes and specimen containers should always be securely capped.
3. The interior of the bowl should be inspected daily for staining or soiling at the level of the rotor. If this is evident then the centrifugation protocols should be re-evaluated.
4. Angle heads should not be used in microbiology except in special, high-speed centrifuges. With ordinary angle heads, some fluid, even from capped tubes, may be ejected because of the geometry of the machine.
5. Except in ultracentrifuges and with small prothrombin tubes, a space of at least 2 cm should be left between the level of the fluid and the rim of the centrifuge tube.

Centrifugation of Risk Group 3 and 4 microorganisms and materials

1. Materials that may contain microorganisms in Risk Group 3 or 4 should be centrifuged in batches separate from other materials.
2. Centrifuge tubes and bottles should have screw caps and should be marked in a way agreed locally to indicate that the contents are in Risk Group 3 or 4.
3. Sealable centrifuge buckets (safety cups) must be used.
4. The buckets must be loaded, sealed and opened in a biological safety cabinet.
5. Buckets, rotors and centrifuge bowls should be regularly decontaminated.

Techniques for the use of homogenizers, shakers and sonicators

1. Domestic (kitchen) homogenizers should not be used in laboratories as they may leak and release aerosols. Laboratory blenders and stomachers are safer (Figs 5 and 6).
2. Caps and cups or bottles should be in good condition and free from flaws or distortion. Caps should be well-fitting and gaskets should be in good condition.
3. Aerosols containing infectious materials may escape from between the cap and the vessel in homogenizers, shakers and sonicators. Pressure builds up in the vessel during operation. Polytetrafluoroethylene (PTFE) vessels are recommended because glass may break, releasing infectious material and possibly wounding the operator.
4. When in use these machines should be covered by a strong transparent plastic casing. This should be disinfected after use. Where possible, the machines, under their plastic covers, should be operated in a biological safety cabinet.

Fig. 5. Waring-type blender for safe blending of hazardous materials
Source: Christison Scientific Equipment Ltd, Gateshead, England.

5. At the end of the operation the container should be opened in a biological safety cabinet.
6. Hearing protection should be provided for people using sonicators.

Techniques for the use of tissue grinders (Griffith's tubes or TenBroek grinders)

1. Glass grinders should be held in a wad of absorbent material in a gloved hand. PTFE grinders are safer.
2. Tissue grinders should be operated and opened in a biological safety cabinet.

41

Fig. 6. Stomacher for blending hazardous materials
Source: Seward Medical, London, England.

Techniques for the care and use of refrigerators and freezers

1. Refrigerators, deep-freezers and dry-ice chests should be defrosted and cleaned periodically and any ampoules, tubes, etc. that have broken during storage removed. Face protection and heavy duty rubber

gloves should be worn. After cleaning, the inner surfaces of the cabinet should be disinfected.

2. All containers stored in refrigerators, etc., should be clearly labelled with the scientific name of the contents, the date stored and the name of the individual who stored them. Unlabelled and obsolete materials should be autoclaved.

3. Flammable solutions must not be stored in a refrigerator unless it is explosion-proof. Notices to this effect should be placed on refrigerator doors.

Techniques for the opening of ampoules containing lyophilized infectious materials

Care should be taken when ampoules of freeze-dried materials are opened as the contents may be under reduced pressure and the sudden inrush of air may disperse some of the contents into the atmosphere. Ampoules should always be opened in a biological safety cabinet.

The following procedures are recommended for opening ampoules:

1. First decontaminate the outer surface of the ampoule.
2. Make a file mark on the tube near to the middle of the cotton or cellulose plug.
3. Hold the ampoule in a wad of cotton to protect the hands.
4. Apply a red-hot glass rod to the file mark to crack the glass.
5. Remove the top gently and treat as contaminated material.
6. If the plug is still above the contents of the ampoule, remove it with sterile forceps.
7. Add liquid for resuspension slowly to the ampoule to avoid frothing.

Storage of ampoules containing infectious materials

Ampoules containing infectious materials should never be immersed in liquid nitrogen because cracked or imperfectly sealed ampoules may break or explode on removal. If very low temperatures are required, ampoules should be stored only in the gaseous phase above the liquid nitrogen. Otherwise, infectious materials should be stored in mechanical deep-freeze cabinets or on solid carbon dioxide (dry ice).

Eye and hand protection should be worn by workers when removing ampoules from cold storage.

The outer surfaces of ampoules stored in these ways should be disinfected when they are removed from storage.

43

Special precautions with blood and other body fluids

The precautions outlined below are designed to protect the worker against infection by bloodborne pathogens, such as hepatitis B virus, human immunodeficiency virus, and the viruses of haemorrhagic fevers, and by protozoa and helminths (*2, 5, 18, 30–34*).

Note. Patients suspected of having haemorrhagic fever may, in fact, have malaria, and a blood film should therefore be examined for malaria parasites before any other laboratory procedure is attempted (see "Blood films", page 45).

Collection, labelling and transport of specimens

1. Gloves must be worn for all procedures.
2. Blood should be collected by experienced staff.
3. After venepuncture, needles should be removed from syringes with needle forceps and placed in special receptacles. The blood should be carefully discharged into the specimen tube and the syringe discarded into the special receptacle. The specimen tube should be securely stoppered.
4. Specimen tubes and request forms should be labelled "Danger of infection" or with a similar or more specific warning.
5. The tubes should be placed in plastic bags for transmission to the laboratory. Request forms should be placed in separate bags or envelopes.
6. Reception staff should **not** open these bags.

Containment

1. Diagnostic work may be done in a Basic laboratory – Biosafety Level 2, preferably one dedicated for this purpose.
2. Research and development work involving propagation or concentration of the microorganisms should be done in a Containment laboratory – Biosafety Level 3.

Opening specimen tubes and sampling contents

1. Specimen tubes should be opened in a Class I or Class II biological safety cabinet.
2. Gloves must be worn.
3. To prevent splashing, the stopper should be grasped through a piece of paper.

Protective clothing

Protective clothing should be supplemented with a plastic apron, gloves, and safety spectacles or visor.

Glass and "sharps"

1. Plastics should replace glass wherever possible. Only tough (borosilicate) glass should be used, and any article that is chipped or cracked should be discarded.
2. Hypodermic needles must not be used. Cannulas are permitted.
3. Scalpels and knives should not be used on unfixed tissues.

Blood films

Blood films should be handled with forceps. Air-dried, thick films from patients with suspected viral haemorrhagic fever should be exposed for 15 minutes to buffered formalin solution (formalin, 380 g/l, 500 ml; sodium dihydrogen phosphate monohydrate ($NaH_2PO_4 \cdot H_2O$), 22.75 g; anhydrous disodium hydrogen phosphate (Na_2HPO_4), 32.5 g; distilled water, 4500 ml), then washed well with tapwater before staining. Thin films should be fixed in methanol for 30 minutes followed by heating at 95°C (2).

Note. Other agents may still be viable.

Automated equipment

1. Equipment should be of the enclosed type.
2. Probes should be shielded to avoid splashing and needle-stick accidents.
3. Effluent should be trapped in closed vessels or discharged at least 25 cm into the waste disposal system.
4. Equipment need not be dedicated if specimens are "batched" and tested together at the end of the session.
5. If the nature of the equipment permits, hypochlorite or glutaraldehyde should be passed through it at the end of each session. Otherwise flushing with water may be adequate.

Centrifuging

1. All centrifuge tubes should be stoppered.
2. Sealed buckets (safety cups) should be used.

Tissues

1. Formalin fixatives should be used. Small specimens, e.g., from needle biopsies, can be fixed and decontaminated within a few hours but larger specimens may take several days.
2. Frozen sectioning should be avoided. If it is essential then the cryostat should be shielded and the operator should wear a visor. Afterwards the temperature of the instrument should be raised to 20°C so that it can be decontaminated.

Decontamination

Hypochlorites and aldehydes are recommended for decontamination. Hypochlorites should contain 1 g/litre available chlorine for general use and 10 g/litre for blood spillages. Glutaral may be used for decontaminating surfaces (see cautionary note on page 62).

Precautions with materials that may contain "unconventional" agents

Unconventional agents—prions or "slow viruses"—are associated with certain transmissible encephalopathies, notably Creutzfeldt-Jakob disease, Gerstmann-Straussler-Scheinker syndrome (GSS) and kuru in humans; scrapie in sheep and goats; bovine spongiform encephalopathy in cattle; and other transmissible encephalopathies of deer, elk and mink.

Although Creutzfeldt-Jakob disease has been transmitted to humans, there appear to be no proven cases of laboratory-associated infections with any of these agents. Nevertheless, it is prudent to observe certain precautions in the handling of material from infected or potentially infected humans and animals.

The material should be handled under Biosafety Level 2 conditions with the following additional precautions, as the agents are not killed by the normal processes of laboratory disinfection and sterilization.

1. Eye protection and gloves should be worn.
2. All manipulations should be conducted in biological safety cabinets.
3. Great care should be exercised to avoid aerosol production and cuts and punctures of the skin.
4. Formalin-fixed tissues should be regarded as still infectious, even after prolonged exposure.
5. Tissue processors should not be used because of the problems of disinfection. Jars and beakers should be used instead.

6. All jars, beakers, washings, wax shavings, instruments, etc. should be collected for decontamination.
7. Such materials should be autoclaved at 134°C for 18 minutes (see page 69 for alternatives).
8. Instruments that cannot be autoclaved should be placed in hypochlorite solution, 10 g/litre available chlorine, and left there for at least 18 hours.
9. Surfaces to be decontaminated should be exposed to hypochlorite, 10 g/litre available chlorine, for at least 30 minutes.

For further information on the handling of unconventional agents see references *2*, *5*, and *35*.

7. Safe shipment of specimens and infectious materials

The safe shipment of diagnostic specimens and infectious substances is the concern of all who are involved in the process – the medical investigator or scientist, the laboratory, and postal and airline personnel. Medical investigators and laboratory workers are anxious that transport procedures do not hinder the examination of a specimen and hence the timely diagnosis of disease. Postal and airline staff are concerned about the possibility of becoming infected as a result of the escape of harmful microorganisms from broken, leaking and improperly packaged material.

Although there are no reports of illnesses attributable to infection during transport of specimens, various and often conflicting safety precautions have been promulgated. The shipment of unidentifiable and unmarked infectious substances is understandably prohibited. While such shipment would entail potential hazards to employees of the transport service, it would represent a far more serious hazard at the receiving laboratory, as packages are often opened by secretarial and other untrained staff. The hazard is compounded by improper packaging, since a broken container may lead to contamination of the environment and infection of personnel. Hand-carriage of infectious substances is strictly prohibited by international air carriers, as is the use of diplomatic pouches.

Some international organizations — the United Nations Committee of Experts on the Transport of Dangerous Goods, the Universal Postal Union (UPU), the International Civil Aviation Organization (ICAO) and the International Air Transport Association (IATA) — have developed guidelines and procedures designed to facilitate the safe and expeditious shipment of infectious substances while at the same time ensuring the safety of transport personnel and the general public. If the system is to be effective it must be understood by all who use it.

These organizations have also developed agreed common definitions, and packaging and labelling requirements (*36*).

Definitions

The definitions adopted for application as from 1991 are as follows.

● *Infectious substances* are substances containing viable microorganisms, including a bacterium, virus, rickettsia, parasite, fungus, or a recombinant, hybrid or mutant, that are known or reasonably believed to cause disease in animals or humans.

This does not include toxins that do not contain any infectious substances. The transport of genetically modified microorganisms is subject to various conditions and the requirements (*36*) should be consulted.

● *Diagnostic specimens* are any human or animal material including, but not limited to, excreta, secreta, blood and its components, tissue and tissue fluids, being shipped for purposes of diagnosis, but excluding live infected animals.

● *Biological products* are either finished biological products for human or veterinary use manufactured in accordance with the requirements of national public health authorities and moving under special approval or licence from such authorities; or finished biological products shipped prior to licensing for development or investigational purposes for use in humans or animals; or products for experimental treatment of animals, and which are manufactured in compliance with the requirements of national public health authorities. They also cover unfinished biological products prepared in accordance with the procedures of specialized government agencies. Live animal and human vaccines are considered to be biological products and not infectious substances.

Some licensed vaccines may present a biohazard only in certain parts of the world. Competent authorities in those places may require these vaccines to comply with the requirements for infectious substances or may impose other restrictions.

Infectious substances and diagnostic specimens likely to contain infectious substances require triple packaging in accordance with the recommendations of the United Nations, IATA and ICAO (*36*; see also below). These regulations and recommendations are subject to periodic revision. Senders are therefore advised to acquaint themselves with current requirements.

Documentation and packaging requirements

There are documentation requirements for the shipment of infectious substances, biologicals and diagnostic specimens (*36*). Information should be obtained from the appropriate national authorities.

Infectious substances and diagnostic material should be packaged in three layers: (*a*) a primary watertight receptacle containing the specimen; (*b*) a secondary watertight receptacle enclosing enough absorptive material between it and the primary receptacle to absorb all of the fluid in the specimen in case of leakage; and (*c*) an outer package which is intended to protect the secondary package from outside influence, such as physical damage and water, while in transit (Fig. 7).

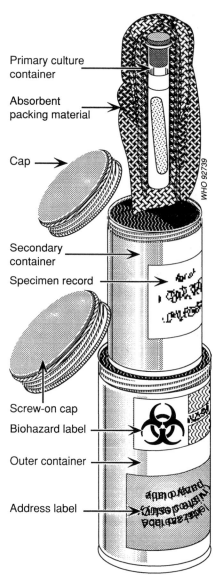

Primary culture container

Absorbent packing material

Cap

Secondary container

Specimen record

Screw-on cap

Biohazard label

Outer container

Address label

Fig. 7. Packing infectious substances for the post

Fig. 8. Infectious substance label

Infectious substances are classed as dangerous goods. Packages containing such substances must bear the infectious substance (biohazard) label (Fig. 8).

One copy of the specimen data forms, letters and other information that identifies or describes the specimen should be taped to the outside of the secondary container. Another copy should be sent by air mail to the receiving laboratory and a third retained by the sender. This enables the receiving laboratory to identify the specimen and decide how to handle and examine it.

If materials are to be consigned in liquid nitrogen or with other protection from ambient or higher temperatures, all containers and packaging should be capable of withstanding very low temperatures and both primary and secondary packaging must be able to withstand a pressure differential of at least 95 kPa and temperatures in the range $-40°C$ to $+50°C$.

If the substance is perishable, warnings should appear on accompanying documents, e.g., "Keep cool, between $+2°C$ and $+4°C$".

Dispatch of packages

Efficient transfer of infectious substances requires good coordination between the sender, carrier and receiving laboratory to ensure that the material is transported safely and arrives on time and in good condition. The following actions are required of the sender:

(a) make advance arrangements with the carrier and the receiver to ensure that the specimens will be accepted and tested promptly; this should be done by telephone and/or cable;

(b) prepare dispatch documents;

(c) arrange routing, by direct flight if possible;

(d) send timely notification of all transportation data to the receiver.

Infectious substances should not be dispatched until advance arrangements have been made between the sender, carrier and receiver, or before the receiver has confirmed with the national authorities that the substances may be imported legally and that no delay will be incurred in the delivery of the package to its destination.

It is the responsibility of the receiver to:

(a) obtain the necessary authorization from the national authorities for the importation of the substance;

(b) provide the sender with the required importation permits, letters of authorization, or other documents required by the national authorities where the specimens originate;

(c) immediately acknowledge receipt to the sender.

Transport-associated accidents: response and emergency safety measures

If packages containing infectious substances are damaged in transit or are thought to be leaking or otherwise faulty, the carriers should contact the sender and consignee of the shipment and the public health authorities. At the same time, the package should be made temporarily safe by the following procedure.

Temporary make-safe procedure for damaged or non-intact packages believed to contain infectious material

1. If broken glass or sharp objects are visible, gather them up with a small dustpan and brush or with forceps, taking care to avoid cutting the hands.
2. Insert hands into a plastic bag to form an improvised mitten.
3. With the hands protected in this way, pick up the package and place it in a plastic bag of suitable size.
4. Discard the improvised mittens into the same bag.

5. Seal the bag and lock it away in a safe place.
6. If any fluid has leaked from the package, disinfect the contaminated area.
7. Wash hands thoroughly.
8. Proceed as shown in the flow diagram (Fig. 9).

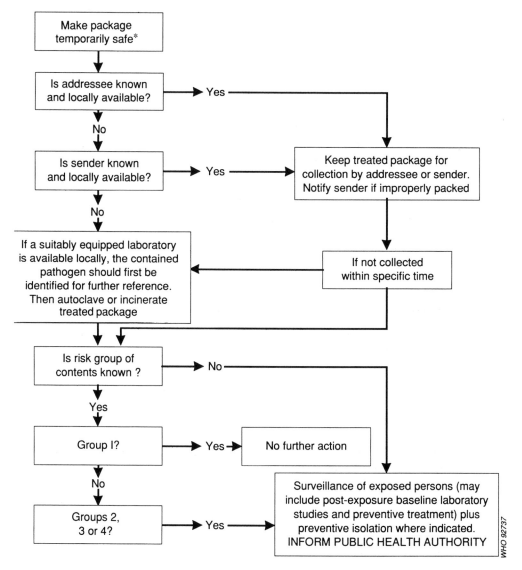

Fig. 9. Guidelines for action in transport-associated accidents to infectious materials in shipment
* See page 52.

The measures that should be taken by public health or veterinary authorities for controlling spread of infection, for disinfection, isolation, administration of vaccines or immunoglobulins, etc., and for notifying the authorities in the other areas where the material may have been handled are similar to those following laboratory accidents and should be provided for in contingency plans (see Chapter 8).

8. Contingency plans and emergency procedures

Every laboratory that works with infective microorganisms should institute safety precautions appropriate to the hazard of the organisms being handled.

A written contingency plan for dealing with laboratory accidents is a necessity in any facility that works with or stores Risk Group 3 or 4 microorganisms (Basic laboratory – Biosafety Level 2, Containment laboratory – Biosafety Level 3 and Maximum Containment laboratory – Biosafety Level 4). As these agents may present a potential hazard to the community, the local health authority should be involved in the development of the plan.

Contingency plans

The contingency plan should provide operational procedures for:

— precautions against natural disasters, e.g., fire, flood, earthquake;
— biohazard risk assessment;
— accident-exposure management and decontamination;
— emergency medical treatment of exposed and injured persons;
— medical surveillance of exposed persons;
— clinical management of exposed persons;
— epidemiological investigation.

In the development of this plan the following items should be considered for inclusion:

— identification of high-risk organisms;
— location of high-risk areas, e.g., laboratories, storage areas;
— identification of at-risk personnel and populations;
— identification of responsible personnel and their duties, e.g., biosafety officer, safety personnel, local health authority, clinicians, microbiologists, veterinarians, epidemiologists, fire and police services;

— lists of clinical treatment and isolation facilities that can receive infected and exposed persons;
— transport of infected and exposed persons;
— lists of sources of immune serum, vaccines, drugs, special equipment and supplies;
— provision of emergency equipment, e.g., protective clothing, disinfectants, decontamination equipment.

General emergency plans for microbiological laboratories

Accidental injection, cuts and abrasions

The affected individual should remove protective clothing, wash the hands and the affected part, apply an appropriate skin disinfectant, go to the first-aid room, and inform the person in charge about the cause of the wound and the organisms involved. If considered necessary, a physician should be consulted and his or her advice followed. Appropriate records should be kept.

Accidental ingestion of potentially hazardous material

Protective clothing should be removed and the individual taken to the first-aid room. A physician should be informed of the material ingested and his or her advice followed. Appropriate records should be kept.

Potentially hazardous aerosol release
(other than in a biological safety cabinet)

All persons should immediately vacate the affected area. The supervisor and the biosafety officer should be informed at once. No-one should enter the room for at least one hour, to allow aerosols to be carried away and heavier particles to settle. Signs should be posted indicating that entry is forbidden. After one hour, decontamination should proceed, supervised by the biosafety officer. Appropriate protective clothing and respiratory protection should be worn for this.

Affected persons should be referred for medical advice.

Broken and spilled cultures

Broken cultures should be covered with a cloth or paper towels. Disinfectant should then be poured over these and left for at least 30 minutes. The cloth or paper towels and the broken material may then be cleared away into a dustpan; glass fragments should be handled with forceps. The contaminated area should then be swabbed with disinfectant. The dustpan containing the broken material, the cloths, towels and swabs should be autoclaved or placed in disinfectant for 24 hours. Gloves should be worn for all these operations.

Spilled cultures should be covered with a cloth or paper towels soaked in disinfectant and left for at least 30 minutes before being mopped up with cloths, which should then be placed in a contaminated-waste container. If request forms or other printed or written matter are contaminated, the information should be copied onto another form and the original discarded into the contaminated-waste container.

Breakage of tubes containing potentially hazardous material in centrifuges not having sealable buckets

If a breakage occurs or is suspected while the machine is running, the motor should be switched off and the machine left closed for 30 minutes. If a breakage is discovered after the machine has stopped the lid should be replaced immediately and left closed for 30 minutes. The biosafety officer should be informed.

Strong (e.g., thick rubber) gloves, covered if necessary with suitable disposable gloves, should be worn for all subsequent operations. Forceps, or cotton held in forceps, should be used to retrieve glass debris.

All broken tubes, glass fragments, buckets, trunnions and the rotor should be placed in noncorrosive disinfectant known to be active against the organisms concerned (not hypochlorite solution, which corrodes metals) and left for 24 hours or autoclaved. Unbroken, capped tubes may be placed in disinfectant in a separate container and recovered after 60 minutes.

The bowl of the centrifuge should be swabbed with the same disinfectant, at the appropriate dilution, left overnight and then swabbed again, washed with water and dried. All swabs should be treated as infected waste.

Breakage of tubes inside sealable buckets (safety cups)

If a breakage is suspected in a sealable bucket it should be opened in a biological safety cabinet. If there has been a breakage, the cap should be replaced but left loose and the bucket autoclaved.

Fire, flood and natural disaster

Fire and other services should be involved in the development of emergency plans. They should be told in advance which rooms contain potentially infectious materials.

After a flood or other natural disaster (including earthquake), local or national emergency services should be warned of the potential hazards within laboratory buildings. They should enter only when accompanied by a trained worker. Cultures and infectious material should be collected in leakproof boxes or strong disposable bags. Salvage or final disposal should be determined by safety staff on the basis of local knowledge.

Vandalism

Vandalism is usually selective, e.g., aimed at animal houses. Suitable defences are strong, heavy doors, good locks and restricted entry. Screened windows and intruder alarms are desirable. Action after vandalism is the same as that for other emergencies.

Emergency services: whom to contact

The telephone numbers and addresses of the following should be prominently displayed near all telephones:

— the institution or laboratory itself (the address and location may not be known in detail by the caller or the services called);
— biosafety officer;
— fire services;
— hospital/ambulance service (if a particular hospital has arranged to accept casualties, e.g., high-risk personnel, the names of individual departments and doctors)
— police;
— medical officer;
— director of the institution or laboratory;
— engineer;
— water, gas and electricity services.

Emergency equipment

The following emergency equipment must be available:

— first aid kit, including universal, cyanide and special antidotes;
— stretcher.

The following are suggested but may be varied according to local circumstances:

— full protective clothing (one-piece coveralls, gloves and head covering – for Risk Group 3 and 4 microorganisms);

58

— full-face respirators with appropriate chemical and particulate filter canisters;

— room disinfection apparatus, e.g., sprays and formaldehyde vaporizers;

— disinfectants;

— tools, e.g., hammers, axes, spanners, screwdrivers, ladders, ropes;

— hazard area demarcation equipment and notices.

Further information is available on medical emergency procedures in outbreaks of infectious disease (31), microbiological laboratory accidents (3, 5, 16), and laboratory first aid procedures (6, 29). Chemical, fire and electrical emergencies are considered in Part 4.

9. Disinfection and sterilization

Chemical disinfectants

There is a considerable diversity in the formulation of proprietary disinfectants. It is therefore essential that manufacturers' recommended use-dilutions are followed.

General precautions

Most disinfectants have some toxic effects (5, 7, 37) (see below and Table 6). Gloves, aprons and eye protection should be worn when bulk disinfectants are being diluted for use.

Chlorine (sodium hypochlorite)

Chlorine is a universal disinfectant active against all microorganisms. It is normally available as sodium hypochlorite, providing various concentrations of available chlorine. It is a strong oxidizing agent and is corrosive to metal. Hypochlorite solutions gradually lose their strength, necessitating daily preparation of fresh dilutions.

A general all-purpose laboratory disinfectant solution should have a concentration of 1 g/litre (1000 ppm) as available chlorine. A stronger solution, containing 5 g/litre (5000 ppm) as available chlorine is recommended for dealing with blood spillage and in the presence of large amounts of organic matter. Sodium hypochlorite containing 5 g/litre (5000 ppm) as available chlorine is recommended as the disinfectant of choice in emergency situations involving viruses such as Lassa virus and Ebola virus.

Many sodium hypochlorite solutions sold for industrial and laboratory use contain 50 g/litre (50 000 ppm) as available chlorine and should therefore be diluted 1:50 or 1:10 to obtain final concentrations of 1 g/litre and 5 g/litre respectively.

Household bleaches usually contain 50 g/litre (50 000 ppm) as available chlorine and should therefore be diluted 1:50 or 1:10 for use.

Calcium hypochlorite granules or tablets contain about 70% available chlorine. Solutions in water may be cloudy. Solutions containing 0.7–1.4 and 7.0 g/litre will contain 500–1000 and 5000 ppm available chlorine respectively.

Sodium dichloroisocyanurate

Sodium dichloroisocyanurate (NaDCC) is available as a powder, containing 60% available chlorine. Solutions containing 0.9–1.7 g/litre and 8.5 g/litre will contain the required amounts of available chlorine. NaDCC is also available as tablets each containing the equivalent of 1.5 g of available chlorine. One or four tablets dissolved in a litre of water will give the required concentrations.

Chloramine

Chloramine powder contains about 25% available chlorine. As it releases chlorine at a slower rate than hypochlorites, a higher initial concentration is required for it to be as effective as hypochlorites. On the other hand,

Table 5. Recommended dilutions of chlorine-releasing compounds

	"Clean" conditions[a]	"Dirty" conditions[b]
Available chlorine required	0.1% (1 g/litre)	0.5% (5 g/litre)
Sodium hypochlorite solution (5% available chlorine)	20 ml/litre	100 ml/litre
Calcium hypochlorite (70% available chlorine)	1.4 g/litre	7.0 g/litre
NaDCC powder (60% available chlorine)	1.7 g/litre	8.5 g/litre
NaDCC tablets (1.5 g available chlorine per tablet)	1 tablet/litre	4 tablets/litre
Chloramine (25% available chlorine)[c]	20 g/litre	20 g/litre

[a] After removal of bulk material.
[b] For flooding, e.g., on blood or before removal of bulk material.
[c] See text.

chloramine solutions are not inactivated by organic matter to the same extent as hypochlorites and a concentration of 20 g / litre is recommended for both "clean" and "dirty" situations.

Table 5 summarizes the recommended dilutions of chlorine-releasing compounds.

Formaldehyde

Formaldehyde is a gas that is active against all microorganisms except at low temperatures, i.e., below 20 °C. It requires a relative humidity of 70%. It is marketed as the solid polymer, paraformaldehyde, in flakes or tablets, or as formalin, a solution of the gas in water of about 370 g / litre (37 %), containing methanol (100 ml/l) as a stabilizer.

Both formulations are used for space decontamination (see page 63). Formaldehyde at 18.5 g / litre (5 % formalin in water) may be used as a liquid disinfectant and is recommended for use against Ebola virus and hepatitis B virus.

Note. Formaldehyde is a suspected carcinogen (7).

Glutaral (glutaraldehyde)

Glutaral is also active against all microorganisms. It is supplied as a solution with a concentration of 20 g / litre (2 %) and most products need to be "activated" before use by the addition of a bicarbonate compound supplied with the pack. This makes the solution alkaline. The activated solution should be used within two weeks (although some commercial solutions remain active for longer periods). Glutaral solutions should be discarded if they become turbid.

Note. Glutaral is now regarded as toxic, irritant and mutagenic (37). Contact with skin, eyes and respiratory tract should be avoided.

Phenolic compounds

Phenolic compounds are active against all vegetative forms of micro-organisms but not against spores. Their activity against nonlipid viruses is variable. A number of common disinfectants are based on phenolic compounds. They may be used when hypochlorites are not available and should be used at the manufacturers' recommended use-dilutions.

Alcohol and alcohol mixtures

Ethanol (ethyl alcohol) and 2-propanol (isopropyl alcohol) have similar disinfectant properties. They are active against vegetative bacteria, fungi and lipid viruses but not against spores. Their action on nonlipid viruses is variable. For highest effectiveness they should be used at concentrations of approximately 70% in water: higher or lower concentrations may not be as germicidal.

Mixtures with other agents are more effective than alcohol alone, e.g., 70% alcohol containing 100 g of formaldehyde per litre, and alcohol containing 2 g/litre (2000 ppm) available chlorine.

Iodine and iodophors

The action of these disinfectants is similar to that of chlorine. Clean surfaces can be treated effectively by solutions containing 0.075 g/litre (75 ppm) as available iodine, but there may be difficulties if an appreciable amount of protein is present.

For hand-washing or as a sporicide, iodophors may be diluted in ethanol. It is suggested that 0.45 g/litre (450 ppm) available iodine is effective against Lassa and Ebola viruses. Polyvidone iodine (PVI) has multiple activity and includes a wetting agent. The common formulation is a 10% solution (1% iodine). For use it may be diluted to 4 times its volume with boiled water. Dilutions should be prepared daily. It should not be used on aluminium or copper.

Hydrogen peroxide

Hydrogen peroxide is a potent disinfectant because of its oxidizing action. It is useful for decontaminating equipment but should not be used on aluminium, copper, zinc or brass. It is supplied as a 30% solution in water and for use is diluted to 5 times its volume with boiled water. It is unstable in hot climates and should always be protected from light.

Table 6 summarizes the properties of these disinfectants. For further information see references 37–39.

Space and surface decontamination

Decontamination of the laboratory space, its furniture and its equipment requires a combination of liquid and gaseous chemical disinfectants.

Contaminated or potentially contaminated surfaces can be decontaminated by a solution of sodium hypochlorite containing 5 g/litre (5000 ppm) as available chlorine or other disinfectant known to be virucidal.

Rooms and equipment can be decontaminated by fumigation with formaldehyde gas, generated by heating paraformaldehyde, 10.8 g/m^3,

Table 6. Properties of some disinfectants[a]

	Active against							Inactivated by					Toxicity		
	Fungi	Bacteria		Myco-bacteria	Spores	Lipid viruses	Nonlipid viruses	Protein	Natural materials	Synthetic materials	Hard water	Detergent	Skin	Eyes	Lungs
		Gram-positive	Gram-negative												
Phenolic compounds	+++	+++	+++	++	−	+	v	+	++	++	+	C	+	+	−
Hypochlorites	+	+++	+++	++	++	+	+	+++	+	+	+	C	+	+	+
Alcohols	−	+++	+++	+++	−	+	v	+	+	+	+	−	−	+	−
Formaldehyde	+++	+++	+++	++++	+++ [b]	+	+	+	+	+	+	−	+	+	+
Glutaral	+++	+++	+++	++++	+++ [c]	+	+	NA	+	+	+	−	+	+	+
Iodophors	+++	+++	+++	+++	+	+	+	+++	+	+	+	A	+	+	−

[a] +++, good; ++, fair; +, slight; −, nil; v, depends on virus; C, cationic; A, anionic; NA, not applicable.
[b] above 40 °C.
[c] above 20 °C.

Source: ref 5. Reproduced by kind permission of the publisher.

or boiling formalin, 35 ml/m^3. All openings in the room (i.e., windows, doors, etc.) should be sealed with masking tape or similar before the gas is generated. Fumigation should be conducted at an ambient temperature of at least 21 °C (70 °F) and a relative humidity of 70%. (See also Decontamination of biological safety cabinets, below.)

Alternatively, to avoid heating, the appropriate amount of paraformaldehyde may be mixed with two parts of potassium permanganate. When water is added the mixture boils violently, generating formaldehyde.

The gas should be in contact with the surfaces to be decontaminated for at least 8 hours. After fumigation the area must be ventilated thoroughly before personnel are allowed to enter. Appropriate respirators must be worn by personnel who have to enter the room before it has been ventilated.

Note. Formaldehyde is a dangerous and irritant gas and is a suspected carcinogen. Full-face respirators with air supply may be necessary. The "two-person" rule should apply.

Decontamination of biological safety cabinets

To decontaminate Class I and Class II cabinets, the appropriate amount of paraformaldehyde should be placed on an electric hot plate (controlled from outside the cabinet). The front closure is sealed into place with masking tape and the heater switched on. The heater should be switched off after one hour or when the paraformaldehyde has all vaporized, and the cabinet should be left undisturbed overnight. The exhaust fan is then switched on and the front closure opened a few millimetres ("cracked"). After a few minutes, the front closure is removed and the fan left on for about an hour, when the formaldehyde will have been exhausted and the cabinet may be used.

Sterilization

Moist steam under pressure is the most effective method of sterilization of laboratory materials. Three types of autoclave are in use: gravity-displacement, vacuum, and fuel-heated pressure cooker.

Gravity displacement autoclaves

Fig. 10 shows the general construction of a gravity displacement autoclave (some models do not have the outer "jacket" which is intended to conserve heat for repeated autoclaving cycles). The autoclave is loaded

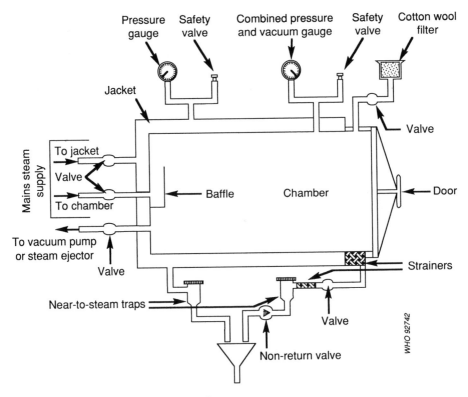

Fig. 10. Gravity displacement autoclave
Source: ref 5. Reproduced by kind permission of the publisher.

and the door closed. Steam from a mains supply (e.g., hospital boiler room) is allowed to enter the chambers through a valve which reduces its pressure. A baffle plate directs the steam to the top of the inner chamber and, as it is less dense than air, it displaces the air downwards through a strainer to remove debris that might block the discharge pipe. There is a trap in this pipe, which is designed to ensure that only saturated steam is retained inside the chamber. The trap opens to allow discharge of air, steam and condensate mixtures if the temperature falls to about 2 °C below that of saturated steam and closes nearer to the temperature of saturated steam. The pressure and hence the temperature in the chamber rise. When the temperature reaches the required level, as indicated by the temperature gauge, that in the load may not have reached the temperature required for proper sterilization, i.e., 121 °C. Time must be allowed, therefore, either determined by previous trials, or indicated by a gauge fitted to a thermocouple placed in the centre of the load. This is the "warming-up" time.

When the load temperature reaches 121 °C, the cycle is allowed to continue for 30 minutes – the "holding time at temperature" (HTAT) –

and then the steam inlet valve is closed. The temperature in the chamber must then be allowed to fall to below 80 °C – the "cooling time" – before the exhaust valve and then the door are opened, or the contents may explode when they are removed.

In some modern gravity displacement autoclaves the entire operation is automatic and may be programmed.

Vacuum autoclaves

Vacuum autoclaves operate automatically according to a predetermined programme. The air in the load is removed by an exhaust pump and steam is admitted and removed in pulses to ensure rapid penetration of the load. The exhaust air may contain infectious aerosols and should therefore be evacuated through a closed system and not directly to the environment. At the end of the cycle, the steam is automatically exhausted and the load is cooled by injection of cold air or a water spray.

These autoclaves can operate at 134 °C and the sterilization cycle may therefore be reduced to 5 minutes.

Fuel-heated pressure cooker autoclaves

These should be used only if a gravity displacement autoclave is not available. A fuel-heated autoclave (Fig. 11) is loaded from the top and heated by gas or electricity. Steam is generated by heating water in the base of the vessel and air is displaced upwards through a relief vent. When all the air has been removed, the valve on the relief vent is closed and the heat reduced. The pressure (and temperature) rise until the safety valve operates at a preset level. This is the start of the holding time. At the end of the cycle the heat is turned off and the temperature allowed to fall to 80 °C or below before the lid is opened.

Loading autoclaves

Materials and objects to be sterilized should be loosely packed in the chamber so that steam can circulate freely and air can be removed easily. Plastic bags should be opened or steam will not penetrate to their contents.

Precautions in the use of autoclaves

There are hazards inherent in the operation of all pressurized vessels. The following rules should be observed.

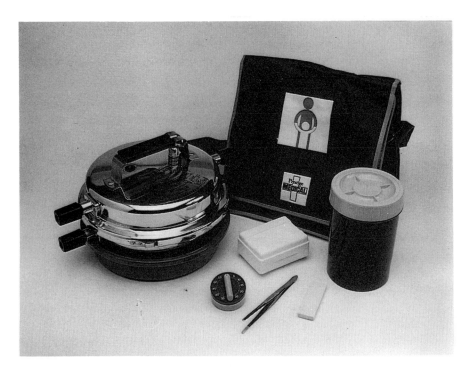

Fig. 11. Portable sterilizer

1. Qualified technicians should regularly inspect the chamber and door seals. A preventive maintenance programme, including a check on gauges and controls, should be carried out at regular intervals.
2. All materials should be in small, shallow containers to aid the removal of air and permit good heat penetration.
3. The chamber should not be tightly packed or heat penetration will be inadequate and some of the load will not be sterilized.
4. If the autoclave is not fitted with an interlocking safety device that prevents the door being opened when the chamber is pressurized, it is essential that the main steam valve is closed and the chamber temperature allowed to fall to below 80 °C before the door is opened. The door should then be opened a few millimetres ("cracked") to allow steam to escape safely, and left in that position for 5 minutes before the autoclave is unloaded.
5. Operators should wear gloves and visors, to protect the arms, hands, face and neck when they open the autoclave, even when the temperature of the contents has been reduced to 80 °C.
6. Biological sterility indicators or thermocouples should be placed at the centre of each load. Regular monitoring with thermocouples and

recording devices in a "worst case" load is highly desirable. Operating cycles can be determined in the light of the findings.

7. Responsibility for operation and routine care should be assigned to trained individuals.

8. The drain screen filter at the bottom of the chamber should be removed and cleaned daily.

9. Care should be taken to ensure that the relief valves of pressure cooker autoclaves do not become blocked by paper, etc. in the load.

Sterilization of "unconventional" agents

Materials suspected of containing unconventional agents or prions (e.g., from Creutzfeldt-Jakob disease, scrapie or bovine spongiform encephalopathy) require high temperatures and/or longer times to be inactivated: 134 °C for 18 minutes HTAT, or six separate cycles at 134 °C with 3 minutes HTAT in each cycle.

Further information on sterilization is given elsewhere (*40–42*).

Incineration

Incineration is a useful method of disposing of laboratory waste either with or without prior autoclaving.

Incineration of infectious materials is an alternative to autoclaving only if:

— the incinerator is under laboratory control;
— it is provided with an efficient means of temperature control and a secondary burning chamber (see below).

Many incinerators, especially those with a single combustion chamber, are inefficient and not satisfactory for dealing with infectious materials, animal carcasses, and plastics. Such materials may not be completely destroyed and the effluent from the chimney may pollute the atmosphere with microorganisms, toxic chemicals and smoke.

There are several satisfactory configurations for combustion chambers but ideally the temperature in the primary chamber should be at least 800 °C and that in the secondary chamber at least 1000 °C. The gas retention time in the secondary chamber should be at least 0.5 seconds.

Materials for incineration, even if they have first been autoclaved, should be transported to the incinerator in bags, preferably plastic. Incinerator attendants should receive proper instructions about loading and temperature control.

Final disposal

The disposal of laboratory and medical waste is subject to various national regulations and WHO has made recommendations on the subject (*39, 43*). In general, ash from incinerators may be treated in the same way as normal domestic waste and removed by local authorities. Autoclaved waste may be disposed of by off-site incineration or in licensed landfill sites.

PART 3

Laboratory equipment

10. Equipment-related hazards

Certain items of equipment (apart from those with unprotected moving parts) may create hazards when they are used: other items are specifically designed to prevent or reduce biological hazards (see Chapter 11).

Equipment that may create a hazard

Table 7 lists equipment and operations that may create hazards and suggests how such hazards may be eliminated or reduced.

Table 7. Equipment and operations that may create hazards

Equipment	Hazard	How to eliminate or reduce the hazard

Hypodermic needles	Accidental inoculation, aerosol or spillage	● Do not recap or clip needles.
		● Use a needle-locking type of syringe to prevent separation of needle and syringe, or use a disposable type where the needle is an integral part of the syringe unit.
		● Use good laboratory techniques, e.g.
		  – Fill the syringe carefully to minimize air bubbles and frothing of inoculum.
		  – Avoid using syringes to mix infectious liquids; if used, ensure that the tip of the needle is held under the surface of the fluid in the vessel and avoid excessive force.
		  – Wrap the needle and stopper in a cotton pledget moistened with an appropriate disinfectant before withdrawing the needle from a rubber-stoppered bottle.

Table 7 (continued)

Equipment	Hazard	How to eliminate or reduce the hazard
Hypodermic needles (continued)		– Expel excess liquid and air bubbles from the syringe vertically into a cotton pledget moistened with an appropriate disinfectant or into a small bottle containing cotton. ● Use a biological safety cabinet for all operations with infectious material. ● Restrain animals while they are being inoculated. Use blunt needles or cannulas for intranasal or oral inoculation. Use a biological safety cabinet. ● Autoclave after use and ensure proper disposal.
Centrifuges	Aerosols, splashing and tube breakage	● Use sealable buckets (safety cups)
Ultra-centrifuges	Aerosols, splashing and tube breakage	● Install HEPA filter between centrifuge and vacuum pump. ● Maintain log book of operating hours for each rotor and a preventive maintenance programme to reduce risk of mechanical failure. ● Load and unload buckets in a biological safety cabinet.
Anaerobic jars	Explosion, dispersing infectious materials	● Ensure integrity of wire capsule around catalyst.
Desiccators	Implosion, dispersing glass fragments and infectious materials	● Place in stout wire cage.
Homogenizers, tissue grinders	Aerosols and leakage	● Operate and open equipment in a biological safety cabinet. ● Use specially designed models that prevent leakage from rotor bearings and O-ring gaskets or use a stomacher. ● Before opening the blender bowl wait 10 minutes to allow the aerosol cloud to settle. Refrigerate to condense aerosols.

Table 7 (continued)

Equipment	Hazard	How to eliminate or reduce the hazard
Sonicators, ultrasonic cleaners	Impaired hearing, dermatitis	● Ensure insulation to protect against subharmonics. ● Wear gloves for protection against high-frequency plus detergent action on skin.
Culture stirrers, shakers, agitators	Aerosols, splashing and spillage	● Operate in a biological safety cabinet or specially designed primary containment. ● Use heavy duty screw-capped culture flasks, fitted with filter-protected outlets, if necessary, and well secured.
Freeze-driers (lyophilizers)	Aerosols and direct contact contamination	● Use O-ring connectors to seal the unit throughout. ● Use air filters to protect vacuum lines. ● Use a satisfactory method of decontamination, e.g., chemical. ● Provide an all-metal moisture trap and a vapour condenser. ● Carefully inspect all glass vacuum vessels for surface scratches. Use only glassware designed for vacuum work.
Domestic-type refrigerators	Provide ignition sources (thermostats, light switches, heater strips, etc.) that can ignite vapours from stored flammable solvents.	● Place warning sign on domestic-type refrigerators: "Do not store flammable solvents in this refrigerator." ● Modify by relocating manual temperature controls to the exterior of the cabinet and sealing all points where wires pass from the refrigerator compartment **Note**: Self-defrosting refrigerators cannot be modified in this way.
Water-baths and Warburg baths	Growth of microorganisms. Sodium azide forms explosive compounds with some metals.	● Regular cleaning and disinfection. ● Do not use sodium azide for preventing growth of organisms.

11. Equipment designed to eliminate or reduce hazards

As aerosols are important sources of infection, care should be taken to reduce the extent of their formation and dispersion. Hazardous aerosols can be generated by many laboratory operations, e.g., blending, mixing, grinding, shaking, stirring and centrifuging of infectious materials (5). Even when safe equipment is used it is best to carry out these operations in an approved biological safety cabinet whenever possible. The use of safety equipment is no assurance of protection unless the operator is trained in its use and uses proper techniques. Equipment should be tested regularly to ensure its continued safe performance.

Table 8 lists safety equipment designed to eliminate or reduce certain hazards and briefly outlines the safety features. Further details of much of this equipment are given in subsequent pages.

Table 8. Safety equipment

Equipment	Hazard corrected	Safety features
Biological safety cabinet		
Class I	Aerosol and spatter	● Minimum inward air flow (face velocity) at work access opening. Adequate filtration of exhaust air (see page 79)
Class II	Aerosol and spatter	● Minimum inward air flow (face velocity) at work access opening. Adequate filtration of exhaust air
Class III	Aerosol and spatter	● Maximum containment
Spatter shield	Spatter of chemicals	● Forms screen between operator and work

Table 8 (continued)

Equipment	Hazard corrected	Safety features
Pipetting aids	Hazards from pipetting by mouth, e.g., ingestion of pathogens, inhalation of aerosols produced by mouth suction on pipette, blowing out of liquid or dripping from pipette, contamination of suction end of pipette	• Ease of use • Control contamination of suction end of pipette, protecting pipetting aid, user and vacuum line • Can be sterilized • Control leakage from pipette tip
Loop micro-incinerators	Spatter from transfer loops	• Shielded in open-ended glass or ceramic tube, heated by gas or electricity
Leakproof vessels for collection and transport of infectious materials for sterilization	Aerosols, spillage and leakage	• Leakproof construction with lid or cover • Durable • Autoclavable
Autoclaves; manual or automatic	Infectious material (made safe for disposal or reuse)	• Approved design • Effective heat sterilization
Screw-capped bottles	Aerosols and spillage	• Effective containment
Vacuum line protection	Contamination of laboratory vacuum system with aerosols and overflow fluids	• Cartridge-type filter prevents passage of aerosols (particle size 0.45 μm) • Overflow flask contains appropriate disinfectant. Rubber bulb may be used to close off vacuum automatically when storage flask is full • Entire unit autoclavable
Goggles or safety spectacles	Impact and splash	• Impact-resistant lenses (must be optically correct or worn over corrective spectacles) • Side shields
Face shield	Impact and splash	• Shields entire face • Easily removable in case of accident

Biological safety cabinets

Biological safety cabinets are the principal items of equipment for providing physical containment. Most laboratory techniques inadvertently produce aerosols. The cabinets act as barriers to minimize the risk of airborne infections by preventing the escape of these aerosols into the laboratory environment and hence their inhalation by the workers. They do not prevent spillage, and are not effective against chemical hazards.

Certain types of cabinet also protect the experiment from airborne contamination.

The selection of a biological safety cabinet is based on the hazard presented by the microorganisms used, the potential of the technique to produce aerosols and the need to protect the work from airborne contamination.

There are three types of biological safety cabinet: Classes I, II and III (1, 5, 21–27). These are described below. Their effectiveness depends on air flow, containment capability, integrity of the high-efficiency particulate air (HEPA) filters, and, in the case of Class I and Class II cabinets,

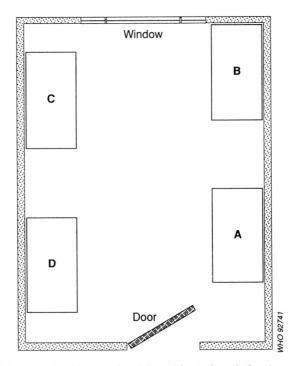

Fig. 12. Possible sites for biological safety cabinets in relation to cross-draughts from door and window and movement of staff. A is bad; B is poor; C is better; D is best
Source: ref 5. Reproduced by kind permission of the publisher.

their position in the room in relation to ventilation currents and the movements of staff. Fig. 12 shows good and bad sites for these cabinets.

Tests for effectiveness should measure: air flow rate and direction, determined with anemometers and smoke generators; ability to retain particles released into the working space (containment factor); and the penetration factor (i.e., filter efficiency) of the HEPA filters. Techniques for these tests are described in various national standards and other publications (21–27).

The cabinets should be tested at the factory, on initial installation in the laboratory and annually thereafter, and whenever they are moved.

Class I biological safety cabinet

A Class I biological safety cabinet (Fig. 13) is an open-fronted, ventilated cabinet for personal protection with an unrecirculated inward air flow away from the operator. It is fitted with a HEPA filter to protect the environment from discharge of microorganisms.

Class I cabinets may be used with low- or moderate-risk micro-organisms (Risk Groups 2 and 3). They provide operator protection but do not protect the material (product) within the cabinet from contamination.

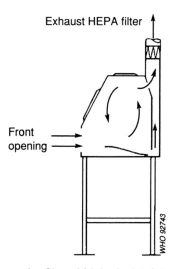

Exhaust HEPA filter

Front opening

WHO 92743

Fig. 13. Schematic diagram of a Class I biological safety cabinet
Source: ref 1. Reproduced by permission of the Minister of Supply and Services Canada.

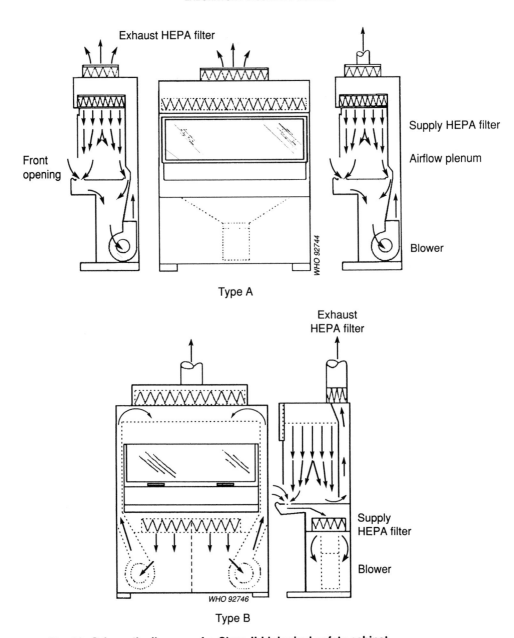

Fig. 14. Schematic diagram of a Class II biological safety cabinet
Source: ref *1*. Reproduced by permission of the Minister of Supply and Services Canada.

Class II biological safety cabinet

A Class II biological safety cabinet (Fig. 14) is an open-fronted, venti-lated cabinet for personal, product and environmental protection, which provides an inward air flow and HEPA-filtered supply and exhaust air.

There are two main variations: the Class IIA type recirculates 70% of the air; the Class IIB type recirculates 30% of the air.

Class IIA cabinets may be used with low- or moderate-risk micro-organisms (Risk Groups 2 and 3), minute quantities of toxic chemicals, and trace amounts of radionuclides. Class IIB cabinets are suitable for larger amounts of toxic, volatile or radioactive substances. Care must be taken in selecting the appropriate Class II cabinet for these purposes.

There are two other kinds of Class II biological safety cabinet: one exhausts all of the air and the other may be converted to other functions.

Class III biological safety cabinet

A Class III biological safety cabinet (Fig. 15) is a totally enclosed, ventilated cabinet which is gas-tight and is maintained under negative air pressure. The supply air is HEPA-filtered and the exhaust air is passed through two HEPA filters in series. Work is performed with attached long-sleeved rubber gloves.

Class III cabinets are used for high-risk microorganisms (Risk Group 4) and provide a total barrier between the operator and the work. Flammable gases should not be used in Class III cabinets.

Class III cabinets may be fitted with dunk tanks for the external disinfection of containers that are passed into or out of the working space. They are often connected in lines by sealable ports. A terminal autoclave may also be included in the line.

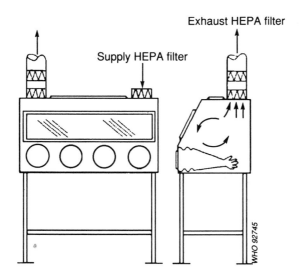

Fig. 15. Schematic diagram of a Class III biological safety cabinet
Source: ref 1. Reproduced by permission of the Minister of Supply and Services Canada.

HEPA filters

Filters suitable for biological safety cabinets (and other microbiological operations) should conform to national standards and ideally not more than three particles should be recovered when the filter is challenged with a dose of 100 000 particles.

Service connections to biological safety cabinets

Gas and electricity cut-offs should be adjacent but external to biological safety cabinets.

Pipetting aids

The importance of pipetting aids cannot be overemphasized. The most common hazards associated with pipetting procedures are the result of mouth suction. Oral aspiration and ingestion have been responsible for many laboratory-associated infections and accidents (5).

Pathogens can also be transferred to the mouth if a contaminated finger is placed on the suction end of a pipette. A lesser-known hazard of mouth pipetting is the inhalation of aerosols caused by suction. The cotton plug is not an efficient microbial filter at negative or positive pressure and particles may be sucked through it. Violent suction may be applied when the plug is tightly packed, resulting in the aspiration of plug, aerosol and even liquid. The ingestion of pathogens is prevented by the use of pipetting aids.

Aerosols may also be generated when a liquid is dropped from a pipette on to a work surface, when cultures are mixed by alternate sucking and blowing, and when the last drop is blown out of a pipette. The inhalation of aerosols unavoidably generated during pipetting operations may be prevented by working in a biological safety cabinet.

Pipetting aids should be selected with care. Their design and use should not create an additional infectious hazard and they should be able to be easily sterilized and cleaned. A selection of these aids is shown in Fig. 16.

Pipettes with cracked or chipped suction ends should not be used as they damage the seating seals of pipetting aids and so create a hazard.

Homogenizers and sonicators

Domestic (kitchen) homogenizers are not sealed and release aerosols. Only homogenizers designed for laboratory use (see Fig. 5, page 41)

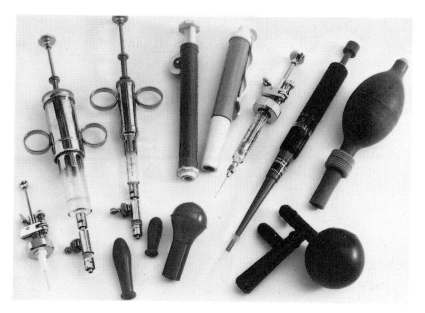

Fig. 16. Pipetting aids
Source: ref 5. Reproduced by kind permission of the publisher.

should be used. Their construction minimizes or prevents such release. The "stomacher" (Fig. 6, page 42) also contains aerosols.

Homogenizers used for Risk Group 3 microorganisms should always be opened in biological safety cabinets.

Sonicators may release aerosols. They should be operated in biological safety cabinets or covered with shields during use. The shields and outsides of sonicators should be decontaminated after use.

Disposable transfer loops

The advantage of disposable transfer loops is that they do not have to be flamed and can therefore be used in biological safety cabinets where Bunsen burners and microincinerators would disturb the air flow. These loops should be placed in disinfectant after use.

Microincinerators

Gas-heated and electrically heated microincinerators (Fig. 17) have borosilicate glass or ceramic shields that minimize the spatter and dispersal of infected material when transfer loops are flamed.

Fig. 17. Gas-heated and electrically heated microincinerators
Source: A R Horwell Ltd, London, England.

PART 4

Chemical, fire and electrical safety

12. Hazardous chemicals

Workers in microbiological laboratories are exposed to chemical hazards as well as to pathogenic microorganisms.

Storage of chemicals

Only minimum amounts of the chemicals listed below should be stored in the laboratory for daily use. Bulk stocks should be kept in specially designated rooms or buildings, which should have concrete floors with sills at doorways to retain spills. Flammable substances should be stored separately in buildings that are some distance from any others. To avoid ignition of flammable and explosive vapours by the sparking of electrical contacts, light switches for these stores should be on the outside of the building and the lights themselves should be in bulkheads.

Chemicals should not be stored in alphabetical order. Otherwise incompatible chemicals (see below) may be in close proximity and some hazardous chemicals may be on high shelves. All large bottles and all bottles containing strong acids and alkalis should be at floor level and in drip trays. Bottle carriers and siphoning devices for filling bottles from bulk containers should be provided. Stepladders should be provided where there are high shelves.

Incompatible chemicals

Many common laboratory chemicals react in a dangerous manner if they come into contact with one another. Some such incompatible chemicals are listed below.

Acetic acid — with chromic acid, nitric acid, hydroxyl compounds, ethylene glycol, perchloric acid, peroxides, and permanganates.

Acetone — with concentrated sulfuric and nitric acid mixtures.

Acetylene — with copper (tubing), halogens, silver, mercury, and their compounds.

Alkali metals, e.g., calcium, potassium and sodium — with water, carbon dioxide, carbon tetrachloride, and other chlorinated hydrocarbons.

Ammonia, anhydrous — with mercury, halogens, calcium hypochlorite and hydrogen fluoride.

Ammonium nitrate — with acids, metallic powders, flammable liquids, chlorates, nitrates, sulfur, and finely divided organic or combustible compounds.

Aniline — with nitric acid and hydrogen peroxide.

Bromine — with ammonia, acetylene, butadiene, butane, hydrogen, sodium carbide, turpentine, and finely divided metals.

Carbon, activated with calcium hypochlorite — with all oxidizing agents.

Chlorates — with ammonium salts, acids, metal powders, sulfur, finely divided organic or combustible compounds, and carbon.

Chlorine — with ammonia, acetylene, butadiene, benzine and other petroleum fractions, hydrogen, sodium carbide, turpentine, and finely divided metals.

Chlorine dioxide — with ammonia, methane, phosphine, hydrogen sulfide.

Chromic acid — with acetic acid, naphthalene, camphor, alcohol, glycerol, turpentine, and other flammable liquids.

Copper — with acetylene, azides, and hydrogen peroxide.

Cyanides — with acids and alkalis.

Flammable liquids — with ammonium nitrate, chromic acid, hydrogen peroxide, nitric acid, sodium peroxide, and halogens.

Hydrocarbons, general — with fluorine, chlorine, formine, chromic acid, sodium peroxide.

Hydrogen peroxide — with chromium, copper, iron, most other metals, their salts, flammable liquids and other combustible products, aniline, and nitromethane.

Hydrogen sulfide — with fuming nitric acid and oxidizing gases.

Iodine — with acetylene and ammonia.

Mercury — with acetylene, fulminic acid, hydrogen.

Nitric acid — with acetic, chromic and hydrocyanic acids, aniline, carbon, hydrogen sulfide, fluids, gases, and other substances that are readily nitrated.

Oxygen — with oils, greases, hydrogen, and flammable liquids, solids and gases.

Oxalic acid — with silver and mercury.

Perchloric acid — with acetic anhydride, bismuth and its alloys, alcohol, paper, wood and other organic materials.

Phosphorus pentoxide — with water.

Potassium permanganate—with glycerol, ethylene glycol, benzaldehyde, and sulfuric acid.

Silver—with acetylene, oxalic acid, tartaric acid, and ammonium compounds.

Sodium—with carbon tetrachloride, carbon dioxide, and water.

Sodium azide—with lead, copper and other metals. This compound is commonly used as a preservative but forms unstable, explosive compounds with metals. If it is flushed down sinks, the metal traps and pipes may explode when worked on by a plumber.

Sodium peroxide—with any oxidizable substance, e.g., methanol, glacial acetic acid, acetic anhydride, benzaldehyde, carbon disulfide, glycerol, ethyl acetate, and furfural.

Sulfuric acid—with chlorates, perchlorates, permanganates, and water.

General rules

Substances in the left-hand column below should be stored and handled so that they cannot accidentally come into contact with the corresponding substances in the right-hand column.

Alkali metals, e.g., sodium, potassium, cerium and lithium	Carbon dioxide, chlorinated hydrocarbons, water
Halogens	Ammonia, acetylene, hydrocarbons
Acetic acid, hydrogen sulfide, aniline, hydrocarbons, sulfuric acid	Oxidizing agents, e.g., chromic acid, nitric acid, peroxides, permanganates

Toxic effects of chemicals

It is now well known that some chemicals adversely affect the health of those who handle them or inhale their vapours. Apart from overt poisons, a number of chemicals are known to have various toxic effects. The respiratory system, blood, lungs, liver, kidneys and the gastrointestinal system, as well as other organs and tissues, may be adversely affected or seriously damaged. Some chemicals are known to be carcinogenic or teratogenic.

Some solvent vapours are toxic when inhaled. Apart from the more serious effects noted above, exposure may result in impairments that show no immediate discernible effects on health, but can include lack of coordination, drowsiness and similar symptoms, leading to an increased proneness to accidents.

Prolonged or repeated exposure to the liquid phase of many organic solvents can result in skin damage. This may be due to a defatting effect, but allergic and corrosive symptoms may also arise.

Many countries publish lists of hazardous chemicals, giving maximum or permissible exposure limits. There are also standard reference works on toxic and hazardous chemicals (*6–8, 29, 44–47*) and on radiation hazards (*29, 47, 48*).

Table 9 lists the reported adverse health effects of some common laboratory chemicals.

Table 9. Adverse health effects of some laboratory chemicals

Chemical	Reported effects	
	Acute	Chronic
Acetaldehyde (acetic aldehyde; ethanal)	Eye and respiratory tract irritation; narcosis	Bronchitis; liver damage
Acetic anhydride (acetyl oxide; ethanoic anhydride)	Strong eye and upper respiratory tract irritation; corrosive action	
Acetone (dimethyl ketone; 2-propanone)	Slight eye, nose and throat irritation; narcosis	
Acetonitrile (methyl cyanide)	Respiratory irritation; cyanide poisoning	
Acrolein	Lacrimation; respiratory irritation	
Ammonia	Eye irritation	Pulmonary oedema
Aniline (amino-benzene; phenylamine)	Cyanosis due to methaemoglobinaemia; slight narcosis; respiratory paralysis	
Benzene	Narcosis	Leukaemia; liver damage; aplastic anaemia
Benzidine	Abdominal pain; nausea; skin irritation	Carcinogenesis
Carbon tetrachloride (tetrachloromethane)	Headache; nausea; slight jaundice; loss of appetite; narcosis	Liver and kidney damage: gastro-intestinal disturbances
Chloroform (trichloromethane)	As for carbon tetrachloride	
Cyanogen bromide	Abdominal pain; nausea; diarrhoea; blurred vision	Pulmonary oedema
Cytochalasin		Mutagenesis
Dioxane	Narcosis	Liver and kidney damage; carcinogenesis
Diethyl ether	Vomiting; eye irritation	Addictive

Table 9 (continued)

Chemical	Reported effects	
	Acute	Chronic
Formaldehyde (formalin)	Respiratory, skin and mucous membrane irritation	Pulmonary oedema
Glutaral	Respiratory and mucous membrane irritation	
Methanol (methyl alcohol)	Narcosis; mucous membrane irritation	Damage to retina and optic nerve
Mercury	Vomiting; diarrhoea; headache; nausea; eye pain	Central nervous system disturbance; swollen gums; loose teeth
α-Naphthylamine		Suspected carcinogen
β-Naphthylamine		Carcinogenesis
Nitrobenzene (nitrobenzol)	Cyanosis due to methaemoglobinaemia; slight narcosis	Anaemia; reduced blood pressure; methaemoglobinaemia with cyanosis; bladder irritation; liver damage
Phenol	Abdominal pain; vomiting; diarrhoea; skin irritation; eye pain; corrosive action	Central nervous system disturbance; coma
Pyridine	Liver and kidney damage	Neurotoxicity
Selenium	Burning skin; eye pain; cough	Central nervous system disturbance; teratogenesis
Tetrahydrofuran (diethyl oxide; tetramethyl oxide)	Narcosis; liver and kidney damage; eye and respiratory irritation	
Thallium	Abdominal pain; vomiting; nausea; diarrhoea	Neuropathy; visual problems; muscle weakness; ataxia
o-Tolidine		Carcinogenesis
Toluene (methyl benzene; phenyl methane; toluol)	Narcosis	Nonspecific neurological impairment; addiction possible
Trichloroethylene (ethinyl trichloride)	Narcosis	Liver damage; nonspecific neurological impairment

Table 9 (continued)

Chemical	Reported effects	
	Acute	Chronic
m-Xylene (1, 2-dimethylbenzene)	Narcosis; headache; dizziness; fatigue; nausea	Nonspecific neurological impairment
o-Xylene (1, 3-dimethylbenzene)	As m-xylene	As m-xylene
p-Xylene (1, 4-dimethylbenzene)	As m-xylene	As m-xylene

Explosive chemicals

Azides should not be allowed to come into contact with copper, e.g., in waste pipes and plumbing. Copper azide will explode violently when subjected to even a mild impact.

Perchloric acid, if allowed to dry on woodwork, brickwork or fabric, will explode and cause a fire on impact.

Picric acid and picrates are detonated by heat and impact.

Chemical spillage

Charts describing methods for dealing with spillages of various chemicals are issued by most manufacturers of laboratory chemicals. Spillage charts and spillage kits are also available commercially. Appropriate charts should be displayed in a prominent position. The following equipment should also be provided:

— protective clothing, e.g., heavy-duty rubber gloves; overshoes or rubber boots;
— scoops and dustpans;
— forceps for picking up broken glass;
— mops, cloths and paper towels;
— buckets;
— soda ash or sodium bicarbonate for neutralizing acids;
— sand;
— nonflammable detergent.

Neutralize spills as follows:

— acids and corrosive chemicals: with soda ash or sodium bicarbonate;
— alkalis: cover with dry sand.

The following actions should be taken in the event of a spillage of a dangerous chemical:

1. Notify the Safety Officer and evacuate non-essential personnel from the area.
2. Attend to persons who may have been contaminated.
3. If the spilled material is flammable, extinguish all naked flames, turn off gas in the room and adjacent areas, and switch off electrical equipment that may spark.
4. Avoid breathing vapour from spilled material.
5. Establish exhaust ventilation if it is safe to do so.
6. Secure the necessary items (see above) to clean up the spillage.

In the event of a large chemical spillage the room should be evacuated and the windows opened, if possible. If the material spilled is flammable, all naked flames in the room concerned and those adjacent should be extinguished and all electrical equipment that may spark switched off.

Compressed and liquefied gases

Rooms where flammable gas cylinders are used should be identified by warning notices on the doors. Not more than one cylinder of a flammable gas should be in the room at any one time. Spare cylinders should be stored in another building at some distance from the laboratory. This store should be locked and identified by a notice.

Compressed gas cylinders should be securely fixed (e.g., chained) to the wall or a solid bench so that they are not dislodged by natural disasters.

Compressed gas cylinders and liquefied gas containers should not be placed near to radiators, naked flames or other heat sources, or sparking electrical equipment, or in direct sunlight.

The main high-pressure valve should be turned off when the equipment is not in use and when the room is unoccupied.

Compressed gas cylinders must be transported with their caps in place and supported on trolleys.

Small, single-use gas cylinders must not be incinerated.

13. Fire in the laboratory

Close cooperation between safety officers and local fire prevention officers is essential. Apart from chemical hazards, the effects of fire on the possible dissemination of infectious material must be considered. This may determine any "burn out" policies (i.e., consideration as to whether it is best to extinguish or contain the fire).

The assistance of local fire prevention officers in the training of laboratory staff in fire prevention, immediate action in case of fire, and the use of fire-fighting equipment is desirable.

Fire warnings, instructions and escape routes should be displayed prominently in each room and in corridors and hallways.

Common causes of fires in laboratories are:

— electrical overloading;
— poor electrical maintenance;
— overlong gas tubing and electricity leads;
— equipment left switched on unnecessarily;
— naked flames;
— deteriorated gas tubing;
— misuse of matches;
— carelessness with flammable materials;
— flammable and explosive chemicals stored in ordinary refrigerators.

Fire-fighting equipment should be placed near to the doors of rooms and at strategic points in corridors and hallways (as advised by local fire prevention officers). This equipment should include hoses, buckets (of water and sand), and the following fire extinguishers: water, carbon dioxide, "dry powder", foam, and bromochlorodifluoromethane (BCF). The shelf-life of these extinguishers should be ascertained and arrangements made for them to be inspected and maintained. Their uses are shown in Table 10.

Further information on the causes and prevention of fires in laboratories is available elsewhere (6, 7, 29, 48).

Table 10. Types and uses of fire extinguishers

Type[a]	Use for	Do not use for
Water	Paper, wood fabric	Electrical fires, flammable liquids, burning metals
CO_2 powder	Flammable liquids and gases, electrical fires	Alkali metals, paper
Dry powder	Flammable liquids and gases, alkali metals, electrical fires	
Foam	Flammable liquids	Electrical fires
BCF	Flammable liquids, electrical fires	

[a] Water extinguishers are driven by CO_2; care is necessary with CO_2 powder extinguishers as the force of the jet may spread burning materials; rooms should be well ventilated after use of BCF extinguishers.

14. Electrical hazards

Electric shock is life-threatening; electrical faults may cause fires. It is therefore essential that all electrical installations and equipment are inspected and tested regularly, including grounding (earthing), and are maintained by qualified electricians. Laboratory staff should not attempt to service any kind of electrical equipment.

Voltages vary from country to country but even low voltages can be hazardous. Care should always be taken to ensure that fuses of the correct rating are interposed between the equipment and the supply. Circuit breakers and ground (earth) fault interrupters should be fitted into laboratory electrical circuits.

Note. Circuit breakers do not protect people; they are intended to protect wiring from overheating and hence to prevent fires. Ground fault interrupters are intended to protect people from electric shock.

All laboratory electrical equipment should be grounded (earthed), preferably through three-prong plugs. Double-insulated devices, requiring only two-prong plugs, are rare in laboratories, but if present may require separate grounding. A ground-free supply may become live as a result of an undetected fault.

All laboratory electrical equipment should conform to a national electrical safety standard (*49*) or that of the International Electrotechnical Commission (*50*).

Laboratory staff should be made aware of the following hazards:

— wet or moist surfaces near to electrical equipment;
— long flexible electrical connecting cables;
— poor and perished insulation on cables;
— overloading of circuits by use of adapters;
— sparking equipment near to flammable substances and vapours;
— electrical equipment left switched on but unattended;
— use of the wrong extinguisher (water or foam instead of CO_2 or BCF) on electrical fires (see Table 10).

More detailed information about electrical hazards is available elsewhere (*6, 8, 29, 50*).

PART 5

Safety organization and training

15. The safety officer and safety committee

It is essential that each laboratory organization has a safety policy, safety code, and a supporting programme for their implementation. The responsibility for this normally rests with the director or head of the institute or laboratory, who may however delegate certain duties to a safety officer or other specialist officers.

It must be emphasized, however, that laboratory safety is also the responsibility of all laboratory employees, and that individual workers are responsible for their own safety and that of their colleagues. Laboratory workers should report unsafe acts or conditions to their superiors.

Periodic safety audits by external independent consultants or specialists are desirable.

Biosafety officer

Wherever possible a biosafety officer should be appointed to ensure that safety policies and programmes are followed throughout the laboratory. The biosafety officer executes these duties on behalf of the head of the institute or laboratory.

In small units, the biosafety officer may be a microbiologist or senior member of the technical staff, who performs these duties on a defined part-time basis. Whatever the degree of involvement in safety work, the person designated should have a sound microbiological laboratory background, be actively involved in the work of the laboratory, and have experience in the broader aspects of laboratory safety. He or she should not be an administrator or technician involved in administrative or clerical activities.

The activities of the biosafety officer should include the following:

(a) periodic internal safety audits on technical methods, chemicals, materials and equipment;

(b) discussions of infringements of the safety code with the appropriate persons;

99

(c) verification that all members of the staff have received appropriate instruction and that they are aware of all hazards, and that members of the medical, scientific and technical staff are competent to handle infectious materials;

(d) provision of continuing instruction in safety for all personnel;

(e) provision of relevant safety literature and information to staff about changes in procedures, technical methods and the introduction of new equipment;

(f) investigation of all accidents and incidents involving the possible escape of potentially infected or toxic material, even if there has been no personal injury or exposure, and reporting of the findings and recommendations to the director and safety committee;

(g) giving assistance in following up sickness or absences among laboratory staff where these may be associated with the work and recorded as possible laboratory-acquired infections;

(h) ensuring that decontamination procedures are followed in the event of a spillage, breakage or other incident involving infectious material. A written record of such accidents and incidents should be kept in case they may be related at a later date to a laboratory-acquired infection or other condition;

(i) ensuring that used materials are decontaminated, and that infectious wastes are safely disposed of after treatment;

(j) ensuring the disinfection of any apparatus requiring repair or servicing before it is handled by non-laboratory personnel;

(k) establishment of procedures for recording the receipt, movements and disposal of pathogenic material and for notification by any research worker or laboratory of the introduction of infectious materials that are new to the laboratory;

(l) advising the director of the presence of any agents that should be notified to the appropriate local or national authorities and regulatory bodies;

(m) reviewing the safety aspects of all plans, protocols and operating procedures for research work;

(n) institution of a system of "on call" duties to deal with any emergencies that arise outside normal working hours.

Biosafety committee

If the institute is large enough, a biosafety committee should be constituted to recommend a safety policy and programme and to formulate or adopt a code of practice or safety manual to serve as the basis of safety practices in the individual laboratories, as advised by the safety officer.

Safety problems brought to the attention of the safety officer, along with information about how they were dealt with, should be presented to the safety committee at regular meetings. Other functions of the committee might include risk assessments of research plans and arbitration in disputes over safety matters.

The size and composition of the safety committee will depend on the size and nature of the laboratory, on the work involved, and on the distribution of its work units or areas. In countries where there is national legislation on health and safety the composition of biosafety committees may already be determined. The composition of a basic safety committee might be:

Chairperson — Elected by members
Members — Biosafety officer
Medical adviser
Representatives of professional staff
Representatives of technical staff
Representatives of management

The biosafety committee might also include in its membership other departmental and specialist safety officers (e.g., with expertise in radiation protection, industrial safety, fire, etc.) and may at times require advice from independent experts in various associated fields, the local authorities, and national regulatory bodies.

General organization

The size and composition of a safety organization will depend on the requirements of the individual laboratory and, in some cases, national regulations. Obviously, small, individual laboratories will not require an elaborate organizational structure or full-time professional safety staff. In many instances, where small individual institutions or laboratories are engaged in work with microorganisms in Risk Groups 1 and 2, a safety organization can be developed for a group of them.

Major biomedical institutions may require a separate biosafety committee which confines its activities to specialized aspects of its own programme.

16. Safety rules for support staff

The safe and optimum operation of a laboratory is dependent to a great extent on the support staff, and it is therefore essential that such personnel should be properly trained for their work.

As engineering, maintenance and cleaning personnel have to enter the premises and interact with the laboratory staff, it is essential that they should perform their duties with due regard for appropriate safety rules. They should apply standard operational procedures, and should be supervised.

Engineering and building maintenance services

These services, which are concerned with the maintenance and repair of the structure, facilities and equipment, have important support functions in the safety programme. For routine servicing and maintenance it is not only convenient but also good safety policy to have available skilled engineers and craftsmen who also have some knowledge of the nature of the work of the laboratory. Safety regulations are better understood and observed by such staff, whereas external engineers and others may be unaware of the hazards and limitations that are placed upon them when they are working in a laboratory, and therefore need much closer supervision by laboratory staff.

The testing of certain items of equipment after servicing is best carried out independently by or under the supervision of the biosafety officer, for example, testing the efficiency of biological safety cabinets after new filters have been fitted, and of other equipment designed to minimize or retain aerosols.

Smaller laboratories or institutions that do not have internal engineering and maintenance services, should establish, well in advance of any emergency, a good relationship with local engineers and builders and familiarize them with the equipment and work of the laboratory.

Engineers and maintenance staff should not enter Containment laboratories – Biosafety Level 3 or Maximum Containment laboratories – Biosafety Level 4, except after clearance by, and under the supervision of, the biosafety officer or the laboratory supervisor.

The staff of large facilities with several laboratories should receive training as outlined on pages 106–108.

Cleaning (domestic) services

Laboratories should preferably be cleaned by staff who are under the control of the laboratory supervisor and not by contract cleaners or staff responsible to other authorities. This practice fosters a safe, cooperative working relationship between the laboratory and cleaning personnel. In particular it ensures that cleaning staff will not be changed without warning.

In Containment laboratories – Biosafety Level 3 and Maximum Containment laboratories – Biosafety Level 4, the cleaning should preferably be done by the laboratory staff. Otherwise cleaning personnel should work only after clearance by, and under the supervision of, the biosafety officer or the laboratory supervisor.

The staff should receive training as outlined on pages 107–108.

The following rules are designed to aid in the prevention of laboratory-acquired infections of cleaning personnel. Copies should be distributed to staff and posted in prominent positions.

Safety rules for domestic and cleaning staff

1. Always wear the protective clothing provided in the manner prescribed by the laboratory supervisor.
2. Take the protective clothing off when you leave a laboratory to visit another part of the building. Do not wear protective clothing when visiting the staff room or canteen.
3. Wash your hands often and always before leaving the laboratory or going to the staff room to eat, drink or smoke.
4. Do not eat, drink, smoke or apply cosmetics in any laboratory. Use the staff room.
5. Do not dust or clean any work benches without the permission of the laboratory staff.
6. If you have an accident of any kind, knock over or break any bottle, tube, jar or piece of equipment, tell the biosafety officer or your supervisor or one of the laboratory staff at once.
7. Do not attempt to clear up after any accident without permission. Do not pick up broken glass with your fingers. Use a dustpan and brush. Follow the instructions of senior members of the staff.
8. Do not enter any room which has a "restricted entry" sign on the door (e.g., the biohazard or radiation hazard sign) unless authorized to do so.
9. Do not empty any discard containers in the laboratory unless a label or an instruction says that you may do so.

---

17. Training programmes

A continuous, on-the-job safety training programme is essential to maintain safety awareness among the laboratory and support staff. Laboratory supervisors, with the assistance of the biosafety officer and other resource persons, play the key role in staff training.

A basic course on good laboratory practice that can be modified to suit most laboratories is offered below. This is followed by five training modules designed for specific members of the laboratory and support staff.

Information on the availability of courses, instructors and visual aids is given in Annex 3.

Basic course: Good laboratory practice

General

1. Sources of laboratory infections.
2. Laboratory hazards:

   (*a*) biological
   (*b*) chemical
   (*c*) physical, including fire and electrical hazards.

3. Workers' rights and duties in relation to safety measures.

Preparatory procedures

1. Access to laboratories.
2. Personal hygiene.
3. Protective clothing.

Experimental procedures

1. Use of mechanical and other pipetting aids.
2. Minimization of aerosol production.

3. Proper use of biological safety cabinets.
4. Proper use of autoclaves and sterilization equipment.
5. Proper use of centrifuges.

Emergency procedures

1. First aid (in laboratories).
2. Spillages and breakages.
3. Accidents.

General laboratory maintenance

1. Storage of hazardous materials.
2. Transport of hazardous materials.
3. Handling and care of laboratory animals.
4. Control of arthropods and rodents.

Check-out procedures

1. Disposal of hazardous waste:

   (*a*)   sterilization
   (*b*)   incineration.

2. Decontamination procedures.
3. Personal hygiene.

Module 1 (the core module): Good microbiological technique

This module is for scientists and technical staff who work in Basic laboratories – Biosafety Levels 1 and 2. The course outlined below can be covered in one week.

As diagnostic laboratories cannot control the kind of specimens they receive and will almost certainly on occasions be required to handle Risk Group 3 microorganisms, some training with these is also necessary.

Course contents

1. Classification of microorganisms, etc., on the basis of hazard; how this is applied in different geographical areas.
2. Laboratory infections; how they occur and the routes and modes of infection.
3. Infections due to known accidents, e.g., accidental inoculation, spillage; prevention or minimization.

4. Infections due to airborne infectious particles; how these particles (aerosols) are released.
5. Measurement and control of aerosols; reduction of hazards by changing techniques and equipment.
6. Protective clothing, face and eye protection, personal hygiene, immunization.
7. Biological safety cabinets, Classes I and II only (Class III cabinets are unlikely to be used in these laboratories).
8. Special precautions for handling blood and body fluids.
9. Disposal of infected laboratory waste; principles and use of autoclaves and incinerators.
10. Chemical disinfection: limitations and policies.
11. The animal house: contamination control.
12. Laboratory design: principles for a safe laboratory.
13. Chemical and carcinogenic hazards; chemical fume cupboards; hazardous analytical equipment.
14. Electrical and fire hazards.
15. First aid for laboratory accidents.
16. Emergency procedures.
17. Mailing and shipping of infectious materials.
18. Examination of existing codes of practice; formulation of local programmes; duties of safety officers; sources of information.
19. How to work without modern facilities.

Module 2: The safe laboratory environment

Module 2 is in two parts, one concerned with planning for safety and the other with organization for safety; both parts are aimed at senior scientific and technical laboratory staff, and at engineering, architectural and administrative staff concerned with construction, maintenance and servicing of the buildings.

Discussion documents to be provided are relevant national biosafety guidelines and laboratory planning and construction codes. Two days should be allowed for this course.

Course contents: Part 1

1. Size and distribution of rooms for different purposes; planning and building systems; furniture and permanent equipment.
2. Services: water, gas, electricity; alternative arrangements where no public services are available.
3. Hygiene facilities: washing, toilets, etc.

4. Ventilation, including that of biological safety cabinets and fume cupboards.
5. Waste disposal; contaminated and chemical waste; autoclaves and incinerators.
6. Animal houses: planning, containment and control; exclusion of unwanted animals including arthropods.
7. Security against vandalism.

Course contents: Part 2

1. Duties and functions of safety committees.
2. Duties and functions of safety officers.
3. Medical supervision; immunization programmes.
4. Training of staff in outlying districts.
5. Consideration of a detailed safety code and how it may be adapted to local circumstances.
6. Safety audits; how they are conducted and what auditors (inspectors) should look for.
7. General safety services, e.g., fire precautions.

Module 3: For support staff

This module, for a one-day course, is for the following groups of laboratory support staff who do not normally have any laboratory training.

Group 1. Domestic staff who: clean the premises; dispose of contaminated and other laboratory waste; wash and prepare glassware and other equipment; prepare and sterilize culture media and reagents.

Group 2. Engineering and maintenance staff who: service laboratory facilities; repair equipment.

Group 3. Staff who: receive and sort pathological material brought to the laboratory; open mail; handle request forms and laboratory records; pack infectious materials for mailing or shipping; drive vehicles that carry infectious materials.

Course contents

1. The nature of microorganisms and how they cause infections (all groups).
2. The work of the laboratory (all groups).
3. How infection may be avoided in the laboratory; personal hygiene, protective clothing, eating, drinking and smoking; biohazard signs and restricted areas (all groups).

4. Use and limitations of disinfectants (all groups).
5. Operation, control and testing of autoclaves and incinerators (groups 1 and 2).
6. Hazards of particular equipment, e.g., biological safety cabinets, incubators, refrigerators, homogenizers and centrifuges (group 2).
7. Hazards of internal transport and of receiving specimens, opening mail and handling records; how infectious materials must be packed for mailing and shipping; emergency action to be taken in case of accident or spillage before the biosafety officer arrives (group 3).
8. Chemical, physical, mechanical and electrical hazards (all groups).
9. The biosafety officer and his or her duties; explanation of the individual worker's rights and responsibilities according to national and local regulations; need to report accidents and unusual occurrences; local security and fire precautions (all groups).
10. Simple first aid (all groups).

Module 4: For safety staff

This module is intended primarily for biosafety officers but other members of the safety committee should be encouraged to attend. Five days should be allowed for this course.

Course contents

1. Outline of legal requirements for conduct in clinical and research laboratories; national occupational health and safety legislation and examples from other, appropriate countries; responsibilities of employers and employees; position of trades unions.
2. Codes of practice and guidelines; the employer's declarations of intent and safety policies; implementation of safety programmes; duties of safety officers and committees.
3. Accident and incident reporting: mechanisms and channels in routine and emergency situations.
4. Emergency programmes; preparing protocols for dealing with accidents, spillage, etc.
5. Medical surveillance: documentation of staff; immunization and sickness records; actions if laboratory infection is suspected.
6. Staff problems leading to aberrant behaviour and consequent hazards.
7. Laboratory and animal house vandalism; security arrangements.
8. Laboratory accidents: inoculation, spillage, breakage, equipment-related hazards: centrifuges, homogenizers, pipetting, microbiological manipulations.

9. Aerosols: how released; equipment- and technique-related; hazards involved; measurement with slit and cascade samplers; theory and practice of HEPA filtration.

10. Supervision and instruction of staff in personal hygiene and use of protective clothing.

11. Biological safety cabinets; classification, limitations and selective use; installation and testing (biological, challenge, chemical smokes); training of users.

12. Design and testing of ventilation systems; graduated pressures; effluent control; clean air rooms.

13. Principles and application of sterilization and disinfection; kinetics of bacterial destruction; autoclaves and their control; use of thermocouples and indicators (chemical and biological).

14. Chemical and gaseous disinfectants; efficiency testing; disinfection policies; irradiation.

15. Packaging, mailing and shipping infectious materials; national and international regulations; emergency procedures.

16. The animal house: containment and control; isolation from the environment.

17. Hazardous chemicals: handling and storage; threshold limit values (or equivalent) and their measurement.

18. Radiation hazards: legal requirements and local control.

19. Life support apparatus; lock-out procedures; the two-person rule.

20. Fire precautions: "burn-out" decisions, i.e., whether to contain or extinguish the fire.

Module 5: For specialist staff who handle microorganisms in Risk Groups 3 and 4

This module is for specialist scientific, technical and safety staff who will handle microorganisms in Risk Groups 3 and 4. Most laboratories will not routinely work with organisms in Risk Group 4, but Risk Group 4 viruses may be encountered in large laboratories. It is essential that safety staff are trained up to the level of Module 4 before following this module. Staff who have not received the proper training should not be given responsibility for microorganisms in Risk Groups 3 and 4. The course lasts two to three days.

Course contents

1. Risk Groups 3 and 4 microorganisms: identity, associated diseases; clinical and epidemiological features; natural reservoirs, vectors; modes of spread.

2. Potential hazards of genetic manipulations and their relation to microorganisms in Risk Groups 3 and 4.

3. Levels of containment for Risk Group 3 and 4 microorganisms.

4. Classes I and II biological safety cabinets. Containment laboratory – Biosafety Level 3 procedures; design of facilities for work with Risk Group 3 microorganisms.

5. Class III biological safety cabinets. Maximum Containment laboratory – Biosafety Level 4 procedures for work with microorganisms in Risk Group 4.

6. Specialized facilities for specific pathogens (e.g., rabies and haemorrhagic fever viruses); monitoring of effluents; specialized clothing and hygiene facilities.

7. Specialized equipment: Class III biological safety cabinets, testing and control, training in use; double-door pass-through autoclaves; warning against full dependence on mechanical equipment that may give a false sense of security.

8. Simple maintenance of electrical, water, pressure equipment, etc. in areas where maintenance staff cannot immediately be admitted.

9. Medical supervision; immunization; emergency measures.

10. Documentation of activities.

PART 6

Safety checklist

18. Safety checklist

This checklist is intended to assist in assessments of the safety status of biomedical laboratories.

Laboratory premises

1. Are the premises generally uncluttered and free from obstructions?
2. Are the premises clean?
3. Are there any structural defects in floors, stairways, walls and roofs?
4. Are floors and stairs uniform and slip-resistant?
5. Are there handrails on flights of stairs with more than four risers?
6. Are there guarded rails on floor openings?
7. Is the working space adequate for safe operation?
8. Are the circulation spaces and corridors adequate for the movement of people and large equipment?
9. Are the benches, furniture and fittings in good condition?
10. Are bench surfaces resistant to solvents and corrosive chemicals?
11. Is there a handbasin in each laboratory room?
12. Are the premises constructed and maintained to prevent entry and harbourage of rodents and arthropods?
13. Are all exposed steam and hot water pipes insulated or guarded to protect personnel?

Storage facilities

1. Are storage facilities, shelves, etc., arranged so that stores are secure against sliding, collapse, falls or spillage?
2. Are storage facilities kept free from accumulations of rubbish, unwanted materials and objects that present hazards from tripping, fire, explosion and harbourage of pests?

Sanitation and staff facilities

1. Are the premises maintained in a clean, orderly and sanitary condition?
2. Is drinking-water available?
3. Are clean and adequate toilet (WC) and washing facilities provided for both male and female staff?
4. Are hot and cold water, soap and towels provided?
5. Are separate changing rooms provided for male and female staff?
6. Is there accommodation (e.g., lockers) for street clothing for individual members of the staff?
7. Is there a staff room for lunch, etc.?
8. Are noise levels acceptable?
9. Is there an adequate organization for the collection and disposal of (noninfectious) rubbish?

Heating and ventilation

1. Is there a comfortable working temperature?
2. Are blinds fitted to windows that are exposed to full sunlight?
3. Is the ventilation adequate, e.g., at least six changes of air per hour, especially in rooms that have mechanical ventilation?
4. Are there HEPA filters in the ventilation system?
5. Does mechanical ventilation compromise air flows in and around biological safety cabinets and fume cupboards?

Lighting

1. Is the general illumination adequate (e.g., 300–400 lux)?
2. Is task (local) lighting provided at work benches?
3. Are there dark or ill-lit corners in rooms and corridors?
4. Are fluorescent lights parallel to the benches?
5. Are fluorescent lights colour-balanced?

Services

1. Is each laboratory room provided with enough sinks, water, electricity and gas outlets for safe working?
2. Is there an adequate inspection and maintenance programme for fuses, lights, cables, pipes, etc?
3. Are faults corrected within a reasonable time?

Security

1. Is the whole building securely locked when unoccupied?
2. Are doors and windows vandal-proof?
3. Are rooms containing hazardous materials and expensive equipment locked when unoccupied?

Fire prevention

1. Is there a fire alarm system?
2. Are all exits unobstructed and unlocked when the building is occupied?
3. Is the fire detection system in good working order and regularly tested?
4. Are the fire doors in good order?
5. Do all exits lead to an open space?
6. Are all exits marked by proper, illuminated signs?
7. Is access to exits marked where the routes to them are not immediately visible?
8. Are any exits obscured by decorations, furniture or equipment?
9. Is access to exits arranged so that it is not necessary to pass through a high-hazard area to escape?
10. Are corridors, aisles and circulation areas clear and unobstructed for movement of staff and fire-fighting equipment?
11. Is all fire-fighting equipment and apparatus easily identified by an appropriate colour code?
12. Are portable fire extinguishers maintained fully charged and in working order, and kept in designated places at all times?
13. Are laboratory rooms with potential fire hazards equipped with appropriate extinguishers for emergency use?
14. If flammable liquids and gases are used in any room, is the mechanical ventilation sufficient to remove vapours before they reach a hazardous concentration?
15. Are fire alarm stations accessible?
16. Are "No smoking" signs posted in areas where smoking is prohibited?

Flammable liquid storage

1. Is the storage facility for bulk flammable liquids separated from the main building?
2. Is it clearly labelled as a fire-risk area?
3. Does it have a gravity or mechanical exhaust ventilation system that is separate from the main building system?

4. Are the switches for lighting sealed or placed outside the building?
5. Are the light fittings inside sealed to protect against ignition of vapours by sparking?
6. Are flammable liquids stored in proper, ventilated containers that are made of non-combustible materials?
7. Are the contents of all containers correctly described on the labels?
8. Are appropriate fire extinguishers placed outside but near to the flammable liquid store?
9. Are "No smoking" signs clearly displayed inside and outside the flammable liquid store?
10. Are only minimum amounts of flammable substances stored in laboratory rooms?
11. Are they stored in properly constructed flammable storage cabinets?
12. Are these cabinets adequately labelled with "Flammable liquid – Fire hazard" signs?

Electrical hazards

1. Are all new electrical installations and all replacements, modifications or repairs made and maintained in accordance with an electrical safety code?
2. Does the interior wiring have a grounded (earthed) conductor (i.e., a three-wire system)?
3. Are circuit breakers and ground fault interrupters fitted to all laboratory circuits?
4. Do all electrical appliances have testing laboratory approval?
5. Are the flexible connecting cables of all equipment as short as practicable, in good condition, and not frayed, damaged or spliced?
6. Is each electric socket outlet used for only one appliance (no adapters to be used)?

Compressed and liquefied gases

1. Is each portable gas container legibly marked with its contents and correctly colour-coded?
2. Are compressed-gas cylinders and their high pressure and reduction valves regularly inspected for safety?
3. Are reduction valves regularly maintained?
4. Is a pressure-relief device connected when a cylinder is in use?
5. Are protection caps in place when cylinders are not in use or are being transported?
6. Are all compressed gas cylinders secured so that they cannot fall, especially in the event of natural disaster?

7. Are cylinders and liquid petroleum gas tanks kept away from sources of heat?

Personal protection

1. Is protective clothing of an approved design provided for all staff for normal work, e.g., gowns, coveralls, aprons, gloves.
2. Is additional protective clothing provided for work with hazardous chemicals and radioactive and carcinogenic substances e.g., rubber aprons and gloves for chemicals and for dealing with spillages; heat-resistant gloves for unloading autoclaves and ovens?
3. Are safety spectacles, goggles and visors provided?
4. Are there eye-wash stations?
5. Are there emergency showers (drench facilities)?
6. Is radiation protection in accordance with national and international standards, including provision of dosimeters?
7. Are respirators available, regularly cleaned, disinfected, inspected and stored in a clean and sanitary condition?

Health and safety of staff

1. Is there an occupational health service?
2. Are first-aid boxes provided at strategic places?
3. Are qualified first-aiders available?
4. Are such first-aiders trained to deal with emergencies peculiar to the laboratory, e.g., contact with corrosive chemicals, accidental ingestion of poisons.
5. Are non-laboratory workers, e.g., domestic and clerical staff, instructed on the potential hazards of the laboratory and the material it handles?
6. Are notices prominently posted giving succinct information about the location of first-aiders, telephone numbers of emergency services, etc.?
7. Are women of childbearing age warned of the consequences of work with certain microorganisms, carcinogens, mutagens and teratogens?
8. Are women of childbearing age told that if they are, or suspect that they are, pregnant they should inform the appropriate member of the medical/scientific staff so that alternative working arrangements may be made for them if necessary?
9. Is there an immunization programme relevant to the work of the laboratory?

10. Are skin tests and/or radiological facilities available for staff who work with tuberculous materials?
11. Are proper records maintained of illnesses and accidents?
12. Are warning and accident prevention signs used to minimize work hazards?

Laboratory equipment

1. Is all equipment certified safe for use?
2. Are procedures available for decontaminating equipment prior to maintenance?
3. Are biological safety cabinets and fume cupboards regularly tested and serviced?
4. Are autoclaves and other pressure vessels regularly inspected?
5. Are centrifuge buckets and rotors regularly inspected?
6. Are hypodermic needles used instead of pipettes?
7. Is cracked and chipped glassware always discarded and not reused?
8. Are there safe receptacles for broken glass?
9. Are plastics used instead of glass where feasible?

Infectious materials

1. Are specimens received in a safe condition?
2. Are specimens unpacked with care and attention to possible breakage and leakage?
3. Are gloves worn for unpacking specimens?
4. Are work benches kept clean and tidy?
5. Do discarded infectious materials, e.g., cultures, accumulate on benches and other places?
6. Are discarded infectious materials removed daily or more often and disposed of safely?
7. Are all members of the staff aware of procedures for dealing with breakage and spillage of cultures and infectious materials?
8. Is the performance of sterilizers checked by chemical, physical and biological indicators?
9. Are centrifuges decontaminated daily?
10. Are sealed buckets provided for centrifuges?
11. Are appropriate disinfectants used correctly?
12. Is there special training for staff who work in Containment laboratories–Biosafety Level 3 and Maximum Containment laboratories–Biosafety Level 4?

Chemicals and radioactive substances

1. Are all chemicals correctly labelled with names and warnings?
2. Are chemical hazard warning charts prominently displayed?
3. Are spillage clearance kits provided?
4. Are staff trained to deal with spillages?
5. Are flammable substances correctly and safely stored in minimal amounts in approved cabinets?
6. Are bottle carriers provided?
7. Is a radiation protection officer available for consultation?
8. Are proper records maintained of stocks and use of radioactive substances?

References

1. Health and Welfare Canada. *Laboratory biosafety guidelines*. Ottawa, Laboratory Centre for Disease Control, 1990.
2. Advisory Committee on Dangerous Pathogens. *Categorization of pathogens and categories of containment*. London, Her Majesty's Stationery Office, 1990.
3. Health Services Advisory Committee. *Safe working and the prevention of infection in clinical laboratories*. London, Her Majesty's Stationery Office, 1991.
4. US Department of Health and Human Services. *National Institutes of Health: Laboratory safety monograph—a supplement to the NIH guidelines for recombinant DNA research*. Washington, DC, 1979.
5. Collins CH. *Laboratory acquired infections: history, incidence, causes and prevention*. 2nd ed. London, Butterworths, 1988.
6. Collins CH, ed. *Safety in clinical and biomedical laboratories*. London, Chapman Hall, 1987.
7. Lieberman DF, Gordon JG, ed. *Biohazards management handbook*. New York, Marcel Dekker, 1989.
8. Pal SB, ed. *Handbook of laboratory safety measures*. 2nd ed. Lancaster, MTP Press, 1990.
9. Drozdov SG et al. [*Essential safety precautions in microbiological and virological laboratories.*] Moscow, Academy of Sciences, 1987 (in Russian).
10. Kent PS, Kubica GP. *A guide to the Level III laboratory*. Atlanta, GA, Centers for Disease Control, 1985.
11. US Department of Health and Human Services. *Biosafety in microbiological and biomedical laboratories*. Washington, DC, Centers for Disease Control and National Institutes of Health, 1988.
12. Canadian Council on Animal Care. *Guide to the care and use of laboratory animals*. Ottawa, Ontario, 1980 (Vol. 1), 1984 (Vol. 2).
13. Council for Accreditation. *Guide for the care and use of laboratory animals*. Joliet, American Association for Accreditation of Laboratory Animal Care, 1987.
14. Melby EC, Altman NH, ed. *CRC handbook of laboratory animal science*. Vol. 1. Cleveland, Ohio, CRC Press, 1974.
15. National Cancer Institute. *Biological hazards in the non-human primate laboratory*. Rockville, MD, Office of Biosafety, 1979.
16. National Research Council. Committee on Hazardous Biological Substances in Laboratories. *Biosafety in the laboratory: prudent practices for the handling and disposal of infectious material*. Washington, National Academic Press, 1989.
17. Iwata K, ed. [*Microbiological biohazards: general considerations and control.*] Tokyo, Soft Science, 1980 (in Japanese).
18. National Committee on Clinical Laboratory Standards (NCCLS). *Protection of laboratory workers from infectious diseases transmitted by blood and tissues. Proposed guidelines 7(9)*. Villanova, NCCLS, 1987 (NCCLS Doc M29-P).
19. Oya A et al., ed. *Handbook for biohazard control*. Tokyo, Kindai Shuppan, 1981.

20. Sterilization and Disinfection Society of Victoria. *Prevention of blood-borne diseases (HBV, NANBH, HIV)*. Parkeville, Victoria, 1989 (Technical Bulletin No 6).
21. British Standards Institution. *Specifications for microbiological safety cabinets.* BS 5726 (Parts 1–4). London, 1991.
22. Canadian Standards Association. *Biological containment cabinets: installation and field testing.* Rexdale, Ontario, 1987 (CAN/CSA Z316.3-M87).
23. Clark RP. *Installation, testing and limitations of microbiological safety cabinets.* Leeds, Science Reviews, 1983.
24. National Sanitation Foundation. *Standard No. 49. Class II (laminar flow) biohazard safety cabinets.* Ann Arbor, 1983.
25. Standards Association of Australia. *Biological safety cabinets (Class 1) for personnel protection.* Sydney, New South Wales, 1991 (AS 2252/1 Part 1).
26. Standards Association of Australia. *Laminar flow biological safety cabinets (Class II) for personnel and product protection.* Sydney, New South Wales, 1991 (AS 2252/2 Part 2).
27. Standards Association of Australia. *Biological safety cabinets: installation and use.* Sydney, New South Wales, 1983 (AS 2647).
28. British Standards Institution. *Specification for safety requirements for laboratory centrifuges.* London, 1980 (BS 4402).
29. Rayburn SR. *The foundations of laboratory safety. A guide for the biomedical laboratory.* New York, Springer Verlag, 1990.
30. Westwood JCN. *The hazard from dangerous exotic diseases.* Philadelphia, PA, Franklin University Press, 1980.
31. Dunsmore DJ. *Safety measures for use in outbreaks of communicable disease.* Geneva, World Health Organization, 1986.
32. United Kingdom Department of Health and Social Security. *Protection against infection with HIV and hepatitis viruses.* London, Her Majesty's Stationery Office, 1990.
33. US Department of Health and Human Services. Recommendations for the prevention of transmission of HIV in health care settings. *Morbidity and mortality weekly reports*, 1987, 36(2S): 35–183.
34. US Department of Health and Human Services. Agent summary statement for HIV, including HTLV III, LAV, HIV1 and HIV2. *Morbidity and mortality weekly reports*, 1988, 37(S4): 1–17.
35. *Guidelines on sterilization and disinfection methods effective against human immunodeficiency virus (HIV)*. 2nd ed. Geneva, World Health Organization, 1989 (WHO AIDS Series No. 2).
36. United Nations. *Recommendations on the transport of dangerous goods.* 7th ed. revised. New York, 1989.
37. New Jersey Department of Health. *Hazardous substances fact sheet: Glutaraldehyde.* CAS No 111–30–8. Philadelphia, 1989.
38. Ayliffe GAJ, Coates D, Hoffman PA. *Chemical disinfection in hospitals.* London, Public Health Laboratory Service, 1984.
39. Suess MJ, Huismans JW. *Management of hazardous waste: policy guidelines and code of practice.* Copenhagen, WHO Regional Office for Europe, 1983 (European Series, No. 14).
40. Block SS. *Disinfection, sterilization and preservation.* 3rd ed. Philadelphia, Lee & Febiger, 1983.
41. Gardner JF, Peel MM. *Introduction to sterilization and disinfection.* Edinburgh, Churchill Livingstone, 1986.
42. Perkins JJ. *Principles and methods of sterilization in health science.* 2nd ed. Springfield, C.C. Thomas, 1982.
43. *Management of waste from hospitals. Report on a WHO meeting.* Copenhagen, WHO Regional Office for Europe, 1985 (Euro Reports and Studies, No. 97).
44. American Chemical Society. *Safety in academic chemistry laboratories.* 3rd ed. Washington, American Chemical Society, 1976.
45. Bretherick L. *Hazards in the chemical laboratory.* London, Butterworths, 1981.

46. National Research Council. Committee on Hazardous Biological Substances in Laboratories. *Biosafety in the laboratory: prudent practices for the handling of hazardous chemicals in the laboratory*. Washington, National Academic Press, 1981.

47. Protection against ionizing radiation. A survey of current world legislation. *International digest of health legislation*, 1972, 22 (4).

48. Furr AK, ed. *CRC handbook of laboratory safety*. 3rd ed. Boca Raton, CRC Press, 1990.

49. United Kingdom Department of Health. *Electrical safety code for hospital laboratory equipment*. London, 1986.

50. International Electrotechnical Commission. *Specifications for safety of electrical equipment used in medical practice*. Geneva, IEC, 1977.

National guidelines and codes of practice

This list includes guidelines and codes of practice for both microbiology and biotechnology.

Australia

Genetic Manipulation Advisory Committee. *Guidelines for small-scale genetic manipulation*. Canberra, 1989.

Standards Association of Australia. *Australian Standard AS 2243, Part 3: Microbiology*. Sydney, New South Wales, 1991.

Bolivia

Instituto Nacional de Laboratorios de Salud. *Medidas de seguridad en los laboratorios de microbiologia, parasitologia o immunologia. [Safe techniques in microbiology, parasitology and immunology laboratories.]* La Paz, 1980.

Canada

Health and Welfare Canada. *Laboratory biosafety guidelines*. Ottawa, Laboratory Centre for Disease Control, 1990.

Chile

Instituto Bacteriologico de Chile. *Manual de bioseguridad.* [*Manual of biosafety.*] Santiago, 1976.

Denmark

Dansk Selskab for Patology. *Om forebyggelse af laboratorie-infektioner.* [*Prevention of laboratory infections.*] Copenhagen, Odontologisk Boghandels Forlag, 1979.

Arbejdstilsynet. *Vurdering af genteknologiske forskningsprojekter.* [*Evaluation of gene technology research projects.*] m.v. Aanvisning Nr 4.6.0.1, Copenhagen, 1990.

Arbejdsmin. *Bekendtgørelse om genteknologi og arbejdsmiljo.* [*Statement on gene technology and the working environment.*] Copenhagen, 1990.

France

Normalisation Française. *Liste des espèces microbiennes communement reconnues comme pathogènes pour l'homme.* [*List of microbial species known to be pathogenic for man.*] Paris, Association Française de Normalisation (AFNOR), 1990.

Germany

Bundesminister für Gesundheit. *Empfehlungen für den Umgang mit pathogenen Mikroorganismen und Klassification von Mikroorganismen und Krankenheitserregern nach den in Umgang mit ihnen auftretenden Gefahren.* [*Recommendations for dealing with pathogenic microorganisms and agents of disease.*] Bonn, Ministry of Health, 1990.

Bundesminister für Forschung und Technologie. *Richtlinien zum Schutz vor Gefahren durch in-vitro neukombinierte Nucleinsauren.* [*Directions for protection against risks during in vitro recombinant DNA work.*] 5th ed. Bonn, 1986.

Hong Kong

Department of Health. *Laboratory safety precautions.* Hong Kong, Queen Mary Hospital, 1990.

India

Department of Biotechnology. *Recombinant DNA safety guidelines and regulations.* Delhi, Ministry of Science and Technology, 1990.

Japan

National Institutes of Health. *Regulations on the safety control of laboratories handling pathogenic agents.* Tokyo, National Institutes of Health, 1987.

Prime Minister. *Guidelines for recombinant DNA experiments.* Tokyo, Ministry of Health, 1987.

Malaysia

Standards and Industrial Research Institute of Malaysia. *Code of practice for safety in laboratories. Part 1, General; Part 2, Chemical; Part 3, Microbiological.* Kuala Lumpur, Ministry of Health, 1986 (MS 19042).

Ministry of Health. *Code of practice for the prevention of infection and accidents in hospital laboratories and post-mortem rooms.* Kuala Lumpur, Ministry of Health, 1986.

Norway

National Institutes of Public Health. [*Safety instructions.*] Oslo, 1990.

Organisation for Economic Co-operation and Development
Recombinant DNA safety considerations. Paris, OECD, 1986.

The Philippines
Republic of the Philippines. *Proposed biosafety guidelines*. Manila, National Committee on Biosafety, 1990.

Republic of Korea
National Institutes of Health. *Studies on the long-term plan and guidelines for genetic engineering research*. Seoul, National Institutes of Health, 1984.

Singapore
Singapore General Hospital. *Laboratory safety manual: diagnostic laboratory*. Singapore, Ministry of Health, 1990.

Sweden
Ministry of Health. *Arbetarskyddsstyrelsens kungörelse med foreskrifter om mikroorganismer samt allmana rad om tillämpningen av foreskrifterna*. [*Statement of the Workers' Protection Board containing regulations on microorganisms, with general advice on their application*.] Stockholm, 1988.

United Kingdom
Advisory Committee on Dangerous Pathogens. *Categorization of pathogens and categories of containment*. London, Her Majesty's Stationery Office, 1990.
Advisory Committee on Genetic Manipulation. *Genetic manipulations regulations*. London, Her Majesty's Stationery Office, 1990.
Health Services Advisory Committee. *Safe working and the prevention of infection in clinical laboratories*. London, Her Majesty's Stationery Office, 1991.

United States of America
US Department of Health, Education and Welfare. National Institutes of Health. *Laboratory safety monograph—a supplement to the NIH guidelines on recombinant DNA research*. Washington, Government Printing Office, 1979.
US Department of Health and Human Services. Guidelines for research involving recombinant DNA molecules. *Federal register*, 51 (88): 7 May 1986.
US Department of Health and Human Services. *Biosafety in microbiological and biomedical laboratories*. Washington, DC, Centers for Disease Control and National Institutes of Health, 1988.

Immunization of staff

1. It is recommended that all laboratory personnel receive protective immunization against the following diseases: diphtheria, hepatitis B, measles, mumps, poliomyelitis, rubella, tetanus, tuberculosis,[a] typhoid fever. Some workers may have been immunized during childhood but documentary evidence of protection should be obtained.

Note. BCG does not appear to give as much protection against tuberculosis in some parts of the world, e.g., the Indian subcontinent, as in others; see *BCG vaccination policies: report of a WHO Study Group*. Geneva, World Health Organization, 1980 (WHO Technical Report Series, No. 652).

2. All persons who work with or who handle animals infected with the following agents should be given the appropriate vaccine or toxoid: *Bacillus anthracis, Clostridium botulinum, Francisella tularensis* type A, *Mycobacterium leprae, Neisseria meningitidis, Yersinia pestis*, louping ill virus,[b] rabies virus, Rift Valley fever virus, Venezuelan equine encephalomyelitis virus, tick-borne encephalitis viruses[b] (Absettarov virus, Hanzalova virus, and Omsk haemorrhagic fever virus).

Advice on sources and use may be obtained from WHO.

---

[a] Caused by *Mycobacterium tuberculosis, M. bovis* and *M. africanum*.

[b] These viruses are antigenically very similar; immunization against one is expected to give protection against others.

Safety in microbiology: training information

Information on the availability of training courses, aids and materials may be obtained by writing to any of the following:

Division of Communicable Diseases, World Health Organization, 1211 Geneva 27, Switzerland.

WHO Collaborating Centre for Biosafety Technology and Consultative Services, Virology Department, Fairfield Hospital, Yarra Bend Road, Fairfield, Victoria 3078, Australia.

WHO Collaborating Centre for Biosafety Technology and Consultative Services, Division of Biosafety, Laboratory Centre for Disease Control, Tunney's Pasture, Ottawa, Ontario, Canada KlA OL2.

WHO Collaborating Centre for Applied Biosafety Programmes and Research, Division of Safety, Building 31, Room 1002, Bethesda, MD 20892, USA.

Office of Biosafety, Centers for Disease Control, 1600 Clifton Road NE, Atlanta, GA 30333, USA.

WHO Collaborating Centre for Biological Safety, National Bacteriological Laboratory, Lundagatan 2, Stockholm, Sweden.

The Division of Communicable Diseases, World Health Organization, can also provide the names and addresses of experts who are willing to travel and conduct courses.

Visual aids

Videotapes and/or slide sets on various aspects of laboratory safety are available from the following:

Learning Resource Activity, Centers for Disease Control, Atlanta, GA 30333, USA.

Commonwealth Scientific and Industrial Research Organization, Film Production Officer, P.O. Box 4, Geelong 3220, Victoria, Australia.

National Safety Council, 425 N. Michigan Avenue, Chicago, IL 60611, USA.

Index

Abrasions   56

Access
    animal facilities   28, 29, 30
    laboratories   9, 20, 24

Accidents (*see also* Spillages)   10, 55–59, 103
    animal-related   29
    transport-associated   52–54

Acetaldehyde   90

Acetic acid   87

Acetic anhydride   90

Acetone   88–90

Acetonitrile   90

Acetylene   88

Acrolein   90

Aerosols
    potentially hazardous releases   56
    prevention   37, 39, 40
    prevention of escape   78, 82
    procedures producing   13, 15

Agitators   75

Air circulation *see* Ventilation

Airlocks   24, 25, 29, 30

Alcohol/alcohol mixtures   62–63, 64

Alkali metals   88

Ammonia   88, 90

Ammonium nitrate   88

Ampoules
    opening   43
    storage   43

Anaerobic jars   74

Aniline   88, 90

Animal facilities   27–31
    Biosafety Level 1   28
    Biosafety Level 2   28–29
    Biosafety Level 3   29
    Biosafety Level 4   30
    containment levels   27
    invertebrates   30–31

Aprons   60, 117

Autoclaves   13, 65–69, 77
    animal facilities   28, 29, 30
    fuel-heated pressure cooker   67, 68
    gravity-displacement   65–67
    laboratories   11, 21
    loading   67
    precautions in use   67–69
    vacuum   67

Autoclaving   16, 17, 47

Azides   92

Basic laboratories   2, 7–18
    chemical, fire, electrical and radiation
        safety   18
    code of practice   7–10
    decontamination and disposal   15–18
    design and facilities   10–12
    equipment   12–13
    health and medical surveillance   13–14
    staff training   14–15, 105–106

Benzene   90

Benzidine   90

Biohazard labels   51

Biohazard warning sign   8, 19, 28

Biological products, definition   49

Biological safety cabinets (BSC)   3, 76, 78–82
    animal facilities   28, 30
    class I   79
    class II   80–81
    class III   81
    containment laboratories   21
    decontamination   65
    recommendations for use   13
    service connections   82
    techniques of use   37–38

Biosafety levels (*see also* Laboratories)   1, 2, 3

Blenders, laboratory   40–41

Blood   15, 38–39
    special precautions   44–46

Blood films 45
Body fluids, special precautions 44–46
Bovine spongiform encephalopathy 46, 69
Bromine 88

Calcium hypochlorite 61
Carbon 88
Carbon tetrachloride 90
Catalase tests 37
Centrifuge buckets, sealed (safety cups) 39, 40
    breakage of tubes in 57
Centrifuging 13, 39–40
    blood and other body fluids 45
    breakage of tubes during 57
    hazards 74
    Risk Group 2 materials 39–40
    Risk Group 3 and 4 materials 40
Checklist, safety 113–119
Chemicals 18, 87–93
    decontamination 16
    explosive 92
    incompatible 87–89
    safety checklist 119
    spillages 92–93
    storage 87–89
    toxic effects 89–92
Chloramine 61–62
Chlorates 88
Chlorine 88
Chlorine dioxide 88
Chlorine-releasing disinfectants 60–62
Chloroform 90
Chromic acid 88
Circuit breakers 96
Cleaning services 103
Clothing, protective 45, 117
    animal house staff 29, 30
    domestic and cleaning staff 103
    emergencies 58
    laboratory staff 9, 19–20, 24
Codes of practice, national 123–125
Containers
    centrifuge 39–40
    leakproof 17, 37, 77
    opening 36, 44
    screw-capped 13, 40, 77
    specimen 35
Containment
    blood specimens 44
    primary 24–25
Containment laboratory 2, 19–23
    code of practice 19–20
    design and facilities 20–21
    equipment 21
    health and medical surveillance 21–23
    support staff 103

Contaminated materials
    autoclaving and recycling 17
    disposal 17–18
    flow chart 18
    identification and separation 16
Contingency plans 55–56
Copper 88
Cosmetics 8, 29, 38
Creutzfeldt-Jakob disease 46, 69
Culture stirrers 75
Cultures, broken and spilled 56–57
Cuts 56
Cyanogen bromide 90
Cylinders, gas 93, 116–117
Cytochalasin 90

Decontamination 9, 15–16
    accidents and spillages 56–57
    biological safety cabinets 65
    blood and other body fluids 46
    space and surface 8, 37, 58, 63–65
    "unconventional agents" 47
Desiccators 74
Design
    basic laboratories 10–12
    containment laboratories 20–21
    maximum containment laboratories
        24–26
Diagnostic specimens 49
Diethyl ether 90
Dioxane 90
Disasters, natural 57–58
Disinfectants, chemical 16, 58, 60–63, 64
Disinfection 60–65
    accidents and spillages 56–57
    waste materials 17
Disposal 12, 15, 16–18, 37, 70
    animal house waste 28, 29, 30, 31
    liquid effluents 21, 24
Documentation, shipment 49–52
Domestic staff
    safety rules 103
    training 107–108

Earthing, electrical equipment 96
Ebola virus 60, 62, 63
Effluents, liquid, disposal 21, 24
Electrical equipment 96
Electrical hazards 18, 96, 116
Electricity services 11, 114
Emergency equipment 58–59
Emergency plans 11, 26, 56–59
Emergency services 58
Engineering staff 102–103, 106, 107

Equipment
    basic laboratories   12–13
    blood and other body fluids   45
    checklist   118
    containment laboratories   21
    electrical   96
    emergency   58–59
    fire-fighting   94
    hazards created by   73–75
    safety   13, 76–84
Eye protection   3, 9, 38, 60, 77, 117

Face protection   9, 38, 77
Fire(s)   18, 94–95
    electrical   96
    emergency plans   57–58
    prevention   115
Fire extinguishers   94, 95, 96, 115
First-aid areas/rooms   11
First-aid kits   58
First-aiders   117
Flammable gases   87, 115
Flammable liquids   87, 115
    incompatible chemicals   88
    storage   43, 115–116
Floods   57–58
Food consumption   8, 29, 38
Footwear   9, 20
Formaldehyde (formalin)   46, 62, 64
    biological safety cabinets   65
    space decontamination   63–65
    toxic effects   91
Freeze driers   75
Freezers   42–43
Fumigation   63–65

Gas supply   12, 114
Gases, compressed and liquefied   93, 116–117
Glass   45
Gloves   3, 9, 29, 44, 57, 60, 117
Glutaral (glutaraldehyde)   46, 62, 64, 91
Good microbiological technique (GMT)   7
    training   105–106
Griffith's tubes   41
Grinders, tissue   41, 74
Ground fault interrupters   96
Grounding, electrical equipment   96
Guidelines, national   123–125

Haemorrhagic fever, viral   44, 45
Hand washing   8, 38, 103
Hand-washing facilities   11, 20, 29
Health effects, adverse, chemicals   89–92
Health surveillance   13–14, 21–23, 117–118
Hearing protection   41
Heating, safety checklist   114

HEPA (high-efficiency particulate air) filters
    21, 24, 29
    biological safety cabinets   79, 80, 81, 82
Hepatitis B virus   44, 62
High-efficiency particulate air filters
    see HEPA filters
Homogenizers   40–41, 74, 82–83
Human immunodeficiency virus   44
Hydrocarbons   88
Hydrogen peroxide   63, 88
Hydrogen sulfide   88
Hypochlorites   46, 47, 60–61, 64

Illumination   11, 114
Immunization   29, 30, 117, 126
Immunocompromised staff   23
Incineration   69
    animal carcasses   29
    waste materials   17–18
Incinerators
    animal facilities   29
    transfer loop   13, 37, 77, 83
Infectious materials, safety checklist   118
Infectious substances
    definition   49
    label   51
Ingestion, accidental   8, 15, 38, 56
Injection, accidental   15, 38, 56
International Air Transport Association
    (IATA)   48, 49
International Civil Aviation Organization
    (ICAO)   48, 49
Invertebrates   30–31
Iodine   63, 88
Iodophors   63, 64

Labels
    blood specimens   44
    "danger of infection"   35, 36, 44
    items for shipment   51
Laboratories
    basic (Biosafety Levels 1 and 2)   7–18
    containment (Biosafety Level 3)   19–23
    maximum containment (Biosafety Level 4)
    24–26
Laboratory premises, safety checklist   113
Laboratory staff
    health and medical surveillance   13–14,
    21–23, 117–118
    personal protection   117
    responsibility for safety   99
    serum samples   10, 23
    training   10, 14–15, 104–110
Laboratory techniques
    safe   35–47
    staff training   106–107

Lassa virus   60, 63
Lighting   11, 114
Loops *see* Transfer loops
Lyophilizers   75

Maintenance staff   102–103, 106–107
Malaria   44
Maximum Containment laboratory   2, 24–26
    design and facilities   24–26
    support staff   103
Medical contact card   22, 23
Medical surveillance   13–14, 21–23, 117–118
Mercury   88, 91
Methanol   91
Microincinerators   37, 77, 83, 84
Microorganisms
    genetically engineered   3, 49
    risk groups   1, 2–3

Naphthylamine   91
Needles, hypodermic   9, 37, 73–74
    disposal   17, 44
    prevention of accidents   38, 45, 73–74
Nitric acid   88
Nitrobenzene   91

Oxalic acid   88
Oxygen   88

Packaging, for shipment   49–52
Perchloric acid   88, 92
Personnel *see* Staff
Phenol   91
Phenolic compounds   16, 62, 64
Phosphorus pentoxide   88
Picric acid/picrates   92
Pipetting   8, 13, 36–37
    aids   13, 36–37, 77, 82, 83
    blood and serum   38
Polyvidone iodine (PVI)   63
Potassium permanganate   89
Pregnant women   14, 23, 117
Premises, laboratory, checklist   113
Prions   46, 69
Pyridine   91

Radiation protection   18, 117
Radioactive substances, safety checklist   119
Refrigerators   42–43, 75
Respirators   20, 58, 65, 117
Risk Group 1   1, 2, 7
    animal facilities   27, 28
    surveillance of workers   14

Risk Group 2   1, 2, 7
    animal facilities   27, 28
    centrifuging techniques   39–40
    surveillance of workers   14
Risk Group 3   2, 3, 19, 20, 21
    animal facilities   27, 29
    centrifuging techniques   40
    staff training   109–110
Risk Group 4   1, 2, 27
    centrifuging techniques   40
    staff training   109–110
Risk groups   1, 2–3

Safety checklist   113–119
Safety committee   100–101, 108
Safety equipment   13, 76–84
Safety officers   99–100, 108
Safety staff   99–101
    training   108–109
Sanitation   21, 114
Scrapie   46, 69
Screw-capped tubes/bottles   13, 40, 77
Security   12, 115
Selenium   91
Serum, separation techniques   38–39
Serum samples, laboratory staff   10, 23
Services   11, 115
    biological safety cabinets   82
Shakers   40–41, 75
Sharps   17, 45
Shipment/transport   35–36, 48–54
    accidents during   52–54
    blood and other body fluids   44
    definitions   49
    dispatch and receipt   51–52
    documentation and packaging   49–52
Silver   89
Skin protection   38
Smoking   8, 29, 38
Sodium   89
Sodium azide   89
Sodium dichloroisocyanurate (NaDCC)   61
Sodium hypochlorite   16, 60–61, 63
Sodium peroxide   89
Sonicators   40–41, 75, 82–83
Specimens   35–36
    blood, precautions   44–46
    collection   44
    containers   35
    diagnostic   49
    labelling   44
    opening   36, 44
    receipt   36, 52
    safety checklist   118
    shipment/transport *see* Shipment/transport

Spillages   10, 56–57
  chemical   92–93
Staff
  facilities   114
  health and safety   117–118
  immunization   29, 30, 117, 126
  laboratory *see* Laboratory staff
  support *see* Support staff
  training *see* Training
Sterilization (*see also* Autoclaves)   24, 65–69
  "unconventional agents"   69
Stirrers, culture   75
Stomachers   40, 42, 83
Storage
  ampoules   43
  chemicals   87–89
  facilities   11, 113
  flammable liquids   43, 115–116
  in freezers and refrigerators   42–43
Sulfuric acid   89
Support staff
  safety rules   102–103
  training   107–108
Syringes   9, 37, 38, 73–74
  disposal   17, 44

TenBroek grinders   41
Tetrahydrofuran   91
Thallium   91
Tissue grinders   41, 74
Tissues, special precautions   46
o-Tolidine   91
Toluene   91

Training   104–110
  animal house staff   30
  information   127–128
  laboratory staff   10, 14–15, 104–110
Transfer loops   37
  disposable   37, 83
Transport, specimen *see* Shipment/transport
Trichloroethylene   91
Two-person rule   19, 30, 65

Ultracentrifuges   74
Ultrasonic cleaners   75
"Unconventional agents"
  precautions   46–47
  sterilization   69
United Nations Committee of Experts on the
    Transport of Dangerous Goods   48, 49
Universal Postal Union (UPU)   48

Vaccines   49, 126
Vacuum line protection   77
Vandalism   12, 58
Ventilation   114
  animal facilities   28, 29, 30, 31
  basic laboratories   11
  containment laboratory   20–21
  maximum containment laboratory   24

Warburg baths   75
Waste disposal *see* Disposal
Water-baths   75
Water supply   11, 21, 114

Xylene   92